McGraw-Hill
My Math

This is your very own math book! You can write in it, draw, circle, and color as you explore the exciting world of math.

Let's get started. Grab a crayon and draw a picture that shows what math means to you.

Have fun!

This is your space to draw.

Mc
Graw
Hill
Education

MHEonline.com

STEM McGraw-Hill is committed to providing
instructional materials in Science, Technology, Engineering,
and Mathematics (STEM) that give all students a solid
foundation, one that prepares them for college and careers
in the 21st century.

Send all inquiries to:
McGraw-Hill Education
8787 Orion Place
Columbus, OH 43240

ISBN: 978-0-07-668791-6 (*Volume 2*)
MHID: 0-07-668791-0

Printed in the United States of America.

4 5 6 7 8 9 10 LMN 22 21 20 19 18 17

Meet The Artists!

Alyssa Gonzalez

King of the Math Jungle This was a great experience similar to being on a roller coaster. Many friends and family supported me in this competition and many kids at Veterans Elementary will benefit from this. *Volume 1*

Finley Moss

Math is Yummy I came up with the idea because my mom and I love to bake cakes and it takes both time and measurement to do so. I was dreaming about how it would feel to win, so I am very excited! Math is yummy after all! *Volume 2*

Other Finalists

Landre Kate Beeler
Pop Numbers 4 Me

Sherry Bergeron's Class*
Math Makes Our World a Better Place

Sally Barmakian's Class*
Math is a Rainbow of Learning

Kamya Cooperwood
Monkey

Judson Upchurch
I like Numbers

Charles "Greyson" Biggs
Rainbow of Possibilities

Leah Rauch
Connect the Dots

Andrew Morris
Math Designed World

Matthew Saldivat
Math Guy in the Sky

Ngun Za Cin
Three-Dimensional Shape Sculpture

Find out more about the winners and other finalists at www.MHEonline.com.

We wish to congratulate all of the entries in the 2011 *McGraw-Hill My Math* "What Math Means To Me" cover art contest. With over 2,400 entries and more than 20,000 community votes cast, the names mentioned above represent the two winners and ten finalists for this grade.

* Please visit mhmymath.com for a complete list of students who contributed to this artwork.

GO digital

it's all at
connectED.mcgraw-hill.com

Go to the Student Center for your eBook, Resources, Homework, and Messages.

Write your Username [_____] Password [_____]

Get your resources online to help you in class and at home.

Vocab

Find activities for building vocabulary.

Watch

Watch animations of key concepts.

Tools

Explore concepts with virtual manipulatives.

Check

Self-assess your progress.

eHelp

Get targeted homework help.

Games

Reinforce with games and apps.

Tutor

See a teacher illustrate examples and problems.

GO mobile

Scan this QR code with your smart phone* or visit mheonline.com/stem_apps.

*May require quick response code reader app.

Available on the App Store

v

Contents in Brief
Organized by Domain

Standards for
Mathematical
PRACTICE → Woven Throughout

Chapter 1
Apply Addition and Subtraction Concepts

Copyright © The McGraw-Hill Companies, Inc. (t)Thinkstock/Comstock/Getty Images, (c)MedioImages/SuperStock, (b)Karin Dreyer/Blend Images/Getty Images

ESSENTIAL QUESTION
What strategies can I use to add and subtract?

Getting Started

I ♥ animals!

Lessons and Homework

Wrap Up

connectED.mcgraw-hill.com

Watch ▶ **Look for this!** Click online and you can watch videos that will help you learn the lessons.

Chapter

2 Number Patterns

ESSENTIAL QUESTION
How can equal groups help me add?

Getting Started

Lessons and Homework

Wrap Up

Welcome to the desert, cowpoke!

Hi!

connectED.mcgraw-hill.com

Chapter 3
Add Two-Digit Numbers

ESSENTIAL QUESTION
How can I add
two-digit numbers?

Getting Started

Lessons and Homework

Wrap Up

We make a great pair!

connectED.mcgraw-hill.com

eHelp
Look for this! Click online and you can get more help while doing your homework.

Chapter 4 Subtract Two-Digit Numbers

ESSENTIAL QUESTION ❓
How can I subtract two-digit numbers?

Getting Started

Lessons and Homework

Wrap Up

Springtime!

connectED.mcgraw-hill.com

Chapter 5
Place Value to 1,000

Number and Operations in Base Ten

ESSENTIAL QUESTION
How can I use place value?

Getting Started

Lessons and Homework

Wrap Up

Look how I've grown!

connectED.mcgraw-hill.com

Tools **Look for this!** Click online and you can find tools that will help you explore concepts.

Chapter 6 Add Three-Digit Numbers

ESSENTIAL QUESTION
How can I add three-digit numbers?

Getting Started

Lessons and Homework

Wrap Up

connectED.mcgraw-hill.com

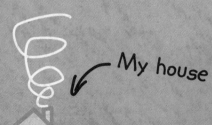

My house

Chapter 7 Subtract Three-Digit Numbers

ESSENTIAL QUESTION
How can I subtract three-digit numbers?

Getting Started

Lessons and Homework

Wrap Up

Look for this!
Click online and you can find activities to help build your vocabulary.

You + School = Cool

connectED.mcgraw-hill.com

Chapter

8 Money

ESSENTIAL QUESTION
How do I count and use money?

Getting Started

Lessons and Homework

Wrap Up

You make me laugh!

Look for this! Click online and you can check your progress.

connectED.mcgraw-hill.com

Chapter

9 Data Analysis

ESSENTIAL QUESTION
How can I record and analyze data?

Go nuts!

connectED.mcgraw-hill.com

Chapter

10 Time

ESSENTIAL QUESTION
How do I use and tell time?

Getting Started

Lessons and Homework

Wrap Up

It's time for our trip!

connectED.mcgraw-hill.com

Chapter 11 Customary and Metric Lengths

Copyright © The McGraw-Hill Companies, Inc.

ESSENTIAL QUESTION
How can I measure objects?

I love rulers because I'm an inch worm!

(tr)Brand X Pictures/Getty Images, (b)Terry Vine/Patrick Lane/Blend Images/Getty Images

connectED.mcgraw-hill.com

Chapter 12 Geometric Shapes and Equal Shares

ESSENTIAL QUESTION
How do I use shapes and equal parts?

Getting Started

Lessons and Homework

Wrap Up

Come on guys. Let's check it out!

connectED.mcgraw-hill.com

Chapter

8 Money

ESSENTIAL QUESTION

How do I count and use money?

Let's Count Money!

Watch a video!

Watch

My Standards

Measurement and Data

2.MD.8 Solve word problems involving dollar bills, quarters, dimes, nickels, and pennies, using $ and ¢ symbols appropriately.

Standards for
Mathematical
PRACTICE ⬇

1. Make sense of problems and persevere in solving them.
2. Reason abstractly and quantitatively.
3. Construct viable arguments and critique the reasoning of others.
4. Model with mathematics.
5. Use appropriate tools strategically.
6. Attend to precision.
7. Look for and make use of structure.
8. Look for and express regularity in repeated reasoning.

= focused on in this chapter

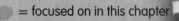

Name ....

Am I Ready?

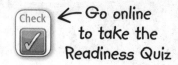 Check ✓

← Go online to take the Readiness Quiz

1. Skip count by 5s.

    5, 10, _____, _____, _____, _____

2. Skip count by 10s.

    10, 20, _____, _____, _____, _____

Count on to add.

3.   25
    + 1

4.   30
    + 2

5.   18
    + 3

6.   16
    + 4

7.   20
    + 5

8.   35
    +10

9. Kim is counting her mittens. She skip counts by 2s. She counts up to 12. How many pairs of mittens does Kim have?

    _____ pairs

How Did I Do? → Shade the boxes to show the problems you answered correctly.

1	2	3	4	5	6	7	8	9

Name _____

My Math Words

Vocab
abc

Review Vocabulary

| equal groups | repeated addition | skip count |

Look at each example. Complete each sentence to describe the example. Then write the repeated addition sentence.

Example	Describe	Write
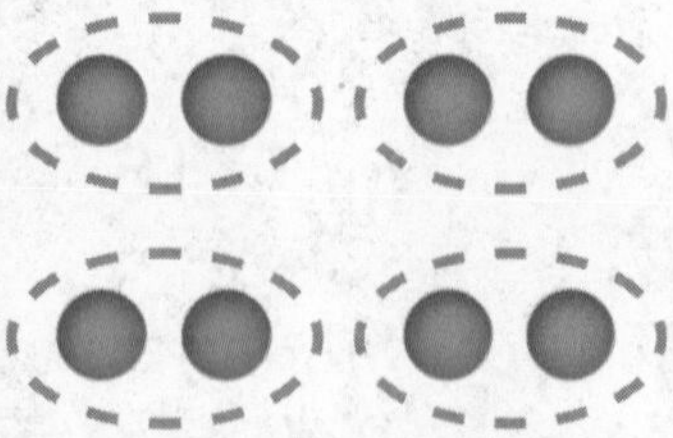	Equal groups of _____. Skip count by _____.	
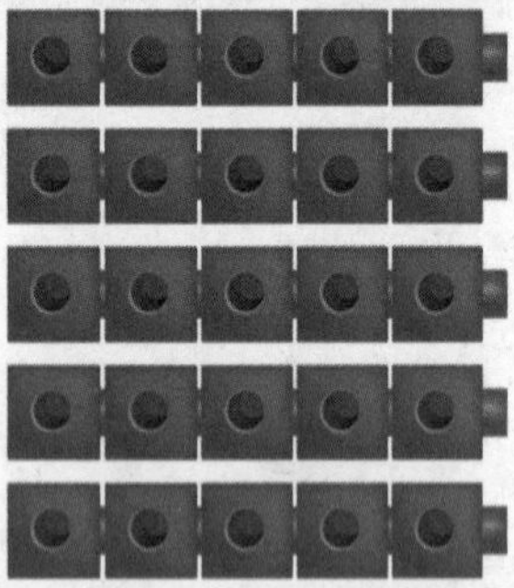	Equal groups of _____. Skip count by _____.	
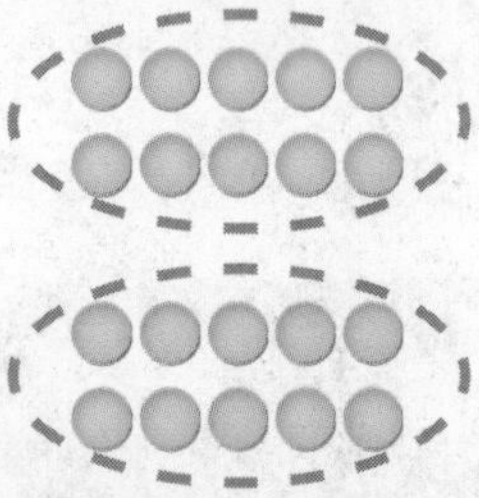	Equal groups of _____. Skip count by _____.	

Lesson 8–1

cent

 → 1 cent

 → 5 cents

Lesson 8–1

cent sign (¢)

1 ¢   5 ¢

Lesson 8–3

dime

 10 cents
10¢

Lesson 8–7

dollar

 100 cents

 100 ¢

Lesson 8–7

dollar sign ($)

$1 or $1.00

Lesson 8–2

nickel

 5 cents
5¢

- Have students think of words that rhyme with some of the words.

- Arrange the cards in alphabetical order.

- Tell students to create riddles for each word. Ask them to work with a friend to guess the word for each word card.

The sign used to show cents.

One penny equals one cent or 1¢. 5 pennies equals 5 cents or 5¢.

One dollar has a value of 100 cents or 100¢.

A dime has a value of 10 cents or 10¢.

A nickel has a value of five cents or 5¢.

The sign used to show dollars.

My Vocabulary Cards

Vocab abc

Mathematical PRACTICE

✂ -

Lesson 8–1

penny

 I cent
I¢

Lesson 8–4

quarter

 25 cents
25¢

Directions:
Ideas for Use

- Have students use a blank card to write this chapter's essential question. Have them use the back of the card to write or draw examples that help them answer the question.

- Use the blank cards to write your own vocabulary cards.

A quarter has a value of 25 cents or 25¢.

A penny has a value of one cent or 1¢.

My Foldable

FOLDABLES Follow the steps on the back to make your Foldable.

78¢

27¢

Juice

50¢

CRAYONS

95¢

①

②

③

Mia

Jasmine

Ella

Ryan

Nathan

Robert

Name

Pennies, Nickels, and Dimes

Explore and Explain
Watch   Tools

pennies
1¢
_____

nickels
5¢
_____

dimes
10¢
_____

 Teacher Directions: Use pennies, nickels, and dimes. Sort the coins. Find the value of each group of coins. Write the value on each bouncy ball machine.

See and Show

Helpful Hint
¢ stands for cents.

dime = 10¢	**nickel** = 5¢	**penny** = 1¢
Count by 10s.	Count by 5s.	Count by 1s.
<u>10</u>¢ <u>20</u>¢	<u>5</u>¢ <u>10</u>¢	<u>1</u>¢ <u>2</u>¢

To find the value of coins, start counting with the coin that has the greatest value.

<u>10</u>¢, <u>20</u>¢, <u>25</u>¢, <u>30</u>¢, <u>31</u>¢, <u>32</u>¢ = <u>32</u>¢

Count to find the value of the coins.

1.

   _____¢, _____¢, _____¢ = _____¢

2.

   _____¢, _____¢, _____¢, _____¢, _____¢, _____¢ = _____¢

Talk Math How many dimes are equal to 70 cents?

Name _____

On My Own

Count to find the value of the coins.

3.

I'm having a ball counting coins!

_____¢, _____¢, _____¢, _____¢, _____¢   = _____¢

4.

_____¢, _____¢, _____¢, _____¢, _____¢, _____¢, _____¢ = _____¢

5.

_____¢, _____¢, _____¢, _____¢, _____¢, _____¢, _____¢ = _____¢

6.

_____¢, _____¢, _____¢, _____¢, _____¢, _____¢, _____¢ = _____¢

Problem Solving

7. Jen had 6 dimes and 4 nickels. She lost 2 of each of them. How much does she still have?

_____ ¢

8. Marcy wants to buy beads that cost 80¢ to make a friendship bracelet. If she has 2 nickels, how many dimes does she need to buy the beads?

_____ dimes

9. Derek has some dimes. He gives Luis 4 dimes. He gives Mia 3 dimes. How much money did Derek give away?

_____ ¢

HOT Problem Paul finds 5 dimes and 2 nickels. He counts them and says he has 50¢. Tell why Paul is wrong. Make it right.

_____

_____

_____

Name _____

My Homework

Homework Helper

 eHelp

Need help? connectED.mcgraw-hill.com

dime = 10¢

Count by 10s.

10¢, 20¢

nickel = 5¢

Count by 5s.

5¢, 10¢

penny = 1¢

Count by 1s.

1¢, 2¢

Helpful Hint
¢ stands for cents.

To find the value of coins, start counting with the coin that has the greatest value.

10¢, 20¢,   25¢, 30¢,   31¢, 32¢ = 32¢

Count to find the value of the coins.

1.

_____¢, _____¢, _____¢, _____¢, _____¢ = _____¢

2.

_____¢, _____¢, _____¢, _____¢, _____¢, _____¢ = _____¢

Count to find the value of the coins.

We're TOPS at finding coin values!

3.

_____¢, _____¢, _____¢, _____¢, _____¢, _____¢ = _____¢

4.

_____¢, _____¢, _____¢, _____¢, _____¢, _____¢ = _____¢

5. Ken has 80¢. His friend has 4 dimes. How many nickels does his friend need to have the same amount of money as Ken?

_____ nickels

Vocabulary Check

Circle the correct answer.

6. **dime**

 Math at Home Have your child count coins to total 90¢.

Name ....

Quarters

Lesson 2
ESSENTIAL QUESTION
How do I count and use money?

Explore and Explain

Tools

MAGIC MONEY COUNTER

quarters 25¢   dimes 10¢   nickels 5¢   pennies 1¢

Teacher Directions: Use quarters, dimes, nickels, and pennies. Sort the coins into the correct columns. Count to find the value of the coins. Write the value on each column.

See and Show

Mathematical
PRACTICE

quarter = 25¢

Count by 25s.

Helpful Hint
Remember ¢ stands
for cents.

<u>25</u>¢, <u>50</u>¢, <u>75</u>¢

Start counting with the coin that has the greatest value.

_____¢, _____¢, _____¢, _____¢, _____¢, _____¢, _____¢ = _____¢

Count to find the value of the coins.

1.

_____¢, _____¢, _____¢, _____¢, _____¢, _____¢ = _____¢

2.

_____¢, _____¢, _____¢, _____¢, _____¢, _____¢ = _____¢

Talk Math   How many quarters do you need
to make 100¢?

Name _____

On My Own

Count to find the value of the coins.

3.

_____¢, _____¢, _____¢, _____¢, _____¢, _____¢, _____¢ = _____¢

4.

_____¢, _____¢, _____¢, _____¢ = _____¢

5.

_____¢, _____¢, _____¢, _____¢, _____¢, _____¢, _____¢ = _____¢

How many quarters do you need to purchase each item?

6.

_____ quarters

7.

_____ quarters

8.

_____ quarters

Problem Solving

Use the information to answer each question.

9. Dale found a quarter, a dime, and three nickels under the sofa. His mom gave him another quarter. Does he have enough money to buy a school basketball game ticket that costs 50¢?

   _____

   Can Dale also buy a juice box for 25¢?

   _____

Shhh!

Dale will never find me here!

10. Lindsay has 2 quarters and 5 dimes. She gives her friend 1 quarter. Lindsay needs 100¢ to buy a stuffed animal. Does she have enough to buy the toy?

    _____

11. Jan has 100¢ in quarters. She wants to buy bracelets. Each bracelet costs a quarter. How many bracelets can she buy?

    _____

HOT Problem Bryan buys water for 75¢. He uses 3 quarters. Describe another way Bryan could have paid for the water.

_____

_____

_____

Name ....

My Homework

Homework Helper
eHelp

Need help? connectED.mcgraw-hill.com

quarter = 25¢

25¢,   50¢,   75¢

Start with the coin that has the greatest value.

25¢,   50¢,   60¢,   70¢,   75¢,   80¢,   81¢   = 81¢

> **Helpful Hint**
> ¢ stands for cents.

Practice

Count to find the value of the coins.

1.

_____¢, _____¢, _____¢, _____¢, _____¢, _____¢   = _____¢

2.

_____¢, _____¢, _____¢, _____¢, _____¢, _____¢, _____¢   = _____¢

Count to find the value of the coins.

3.

_____¢, _____¢, _____¢, _____¢, _____¢, _____¢ = _____¢

Circle the correct number of quarters.

4. Jamal wants to donate 75¢ to the animal shelter. How many quarters would that be?

We love Jamal!

---

5. Jeff has 3 quarters. His friend has 2 quarters. How many more cents does Jeff have than his friend?

_____¢

Vocabulary Check

Circle the correct answer.

6. **quarter**

Math at Home Have your child use quarters to show you 50¢ and 75¢.

Name ....

Count Coins

Lesson 3

ESSENTIAL QUESTION
How do I count and
use money?

Explore and Explain

 Tools

Quarters 25¢   Dimes 10¢   Nickels 5¢   Pennies 1¢

The value of all of the coins is _____.

Teacher Directions: Use quarters, dimes, nickels, and pennies. Sort the coins
into the appropriate columns. Trace them. Write the total value of the coins.

See and Show

Mathematical PRACTICE Skip count!

To count a group of coins, start with the coin that has the greatest value. Count to find the total.

(25¢) (25¢) (10¢) (1¢) (1¢)

25¢, **50**¢, **60**¢, **61**¢, **62**¢

= **62**¢

Count to find the value of the coins.

1.

(10¢) (10¢) (5¢) (5¢) (5¢) (1¢)

_____¢, _____¢, _____¢, _____¢, _____¢, _____¢

= _____¢

2.

(25¢) (10¢) (10¢) (10¢) (5¢) (1¢)

_____¢, _____¢, _____¢, _____¢, _____¢, _____¢

= _____¢

Talk Math How does skip counting help you count groups of different coins?

On My Own

Count to find the value of the coins.

3.

_____¢, _____¢, _____¢, _____¢, _____¢, _____¢

= _____¢

4.

_____¢, _____¢, _____¢, _____¢, _____¢, _____¢

= _____¢

Draw and label the coins from greatest to least.
Find the value of the coins.

5.

= _____¢

Problem Solving

6. Suppose you have 1 quarter, 3 dimes, 1 nickel, and 7 pennies. How much money do you have?

_____¢

7. Luke wants to buy a bouncy ball that costs 25 cents. He has five pennies, 1 dime, and 2 nickels. Does Luke have enough money?

_____

8. Connor has a quarter and a nickel. He gets 2 more quarters for helping around the house. How much money does he have now?

_____¢

Write Math   Chase has 5 dimes. Dan has 10 nickels. Who has more money? Explain.

_____

_____

_____

Name _____

My Homework

Lesson 3

Count Coins

Homework Helper Need help? connectED.mcgraw-hill.com

To count coins, start with the coin that has the greatest value. Count to find the total value.

(25¢) (25¢) (10¢) (5¢) (5¢) (1¢)

25¢,   50¢,   60¢,   65¢,   70¢,   71¢

= 71¢

Count to find the value of the coins.

1.

(25¢) (10¢) (10¢) (5¢) (5¢) (1¢)

_____¢, _____¢, _____¢, _____¢, _____¢, _____¢

= _____¢

2.

(25¢) (25¢) (10¢) (10¢) (5¢) (5¢) (5¢)

_____¢, _____¢, _____¢, _____¢, _____¢, _____¢, _____¢

= _____¢

Count to find the value of the coins.

3.

_____¢, _____¢, _____¢, _____¢, _____¢

= _____¢

4.

_____¢, _____¢, _____¢, _____¢, _____¢, _____¢

= _____¢

5. Kate has 6 dimes, 5 nickels and 4 pennies. How much money does Kate have?

_____¢

Test Practice

6. Find the value of the coins.

41¢   46¢   51¢   36¢
○   ○   ○   ○

I hope Kate has enough money to buy me a cool toy!

Math at Home Give your child coins with a value under $1.00 and have him or her practice counting the coins. Then pretend you are buying and selling things using the coins.

Check My Progress

Vocabulary Check

| penny | nickel | dime | quarter |

Complete each sentence.

1. A coin that has a value of 25 cents is a _____.

2. A coin that has a value of 5 cents is a _____.

3. A coin that has a value of 1 cent is a _____.

4. A coin that has a value of 10 cents is a _____.

Concept Check

Count to find the value of the coins.

5.

_____¢, _____¢, _____¢, _____¢, _____¢, _____¢, _____¢ = _____¢

6.

_____¢, _____¢, _____¢, _____¢, _____¢, _____¢, _____¢ = _____¢

Count to find the value of the coins.

7.

_____ ¢, _____ ¢, _____ ¢, _____ ¢, _____ ¢, _____ ¢   = _____ ¢

8.

_____ ¢, _____ ¢, _____ ¢, _____ ¢, _____ ¢, _____ ¢, _____ ¢, _____ ¢

= _____ ¢

9.

_____ ¢, _____ ¢, _____ ¢, _____ ¢, _____ ¢, _____ ¢

= _____ ¢

Test Practice

10. Myla needs 55¢ to buy a bag of popcorn.
Which coins should she use?

    quarter, quarter   quarter, penny, nickel
       ○   ○

  quarter, quarter, nickel   quarter, quarter, penny
       ○   ○

Measurement and Data
2.MD.8

Problem Solving

STRATEGY: Act It Out

Gavin has 2 quarters, 1 dime, [Watch] [Tools] and 1 nickel. Does he have enough money to buy this toy?

Yippee, Gavin wants me!

65¢

1 Understand
Underline what you know.
Circle what you need to find.

2 Plan
How will I solve the problem?

3 Solve
Act it out.

(25¢) (25¢) (10¢) (5¢)

25¢,   50¢,   60¢,   65¢   = __65__ ¢

Does Gavin have enough money to buy the toy?

yes

4 Check
Is my answer reasonable? Explain.

Practice the Strategy

Robin wants to buy 3 rings.
Each ring costs 20¢. She has
1 quarter, 2 dimes, and 5 nickels.
Does she have enough money?

I need more fingers!

1 Understand Underline what you know.
Circle what you need to find.

2 Plan How will I solve the problem?

3 Solve I will...

4 Check Is my answer reasonable? Explain.

Name _____

Apply the Strategy

Your quarter's safe with me!

1. Maria has 1 quarter in her piggy bank. Her mom gives her a nickel. Her dad gives her a dime. How much money does Maria have in all?

   _____

2. Mark has 2 quarters, 1 dime, and 1 penny. He wants to buy a toy truck for 55¢. Does he have enough money to buy the toy truck?

   _____

3. Wesley has 2 quarters, 3 dimes, and 2 nickels. He has enough money to buy a race car. What is the greatest amount of money that the race car could cost?

   _____

Review the Strategies

Choose a strategy
- Act it out.
- Draw a picture.
- Use logical reasoning.

4. Riley has 1 dime, 3 nickels, and 4 pennies. Does she have enough to buy a cookie that costs 30¢?

   _____

   How much more does she need?

   _____

5. Annie has coins to buy a gel pen at the store. It costs 85¢. She has 2 quarters and 1 nickel. What two coins does she still need?

gel pen
85¢

   _____

6. A notebook costs 40¢. What three coins could you use to pay for the toy?

   _____

Name
....

Dollars

Explore and Explain 🛠 Tools

Hi! I'm George.

one dollar = 100 cents

_____   _____   _____   _____

 Teacher Directions: Count pennies to 100¢. Write the number of quarters it takes to equal 100¢. Do the same for the dime and each of the other coins.

See and Show

Dollar Sign
→ $1.00 ←
Decimal Point

One **dollar** has a value of 100 cents or 100¢. To write one dollar, use a. **dollar sign.**

Use a decimal point to separate the dollars from the cents.

one dollar bill = $1.00

100 pennies = $1	20 nickels = $1	10 dimes = $1	4 quarters = $1

Count to find the value of the coins.
Circle the combinations that equal $1.00.

1.

_____

2.

_____

Talk Math How are $ and ¢ different? How are they alike?

Name _____

On My Own

Count to find the value of the coins.
Circle the combinations that equal $1.00.

3.

_____

4.

_____

5.

_____

6.

_____

7.

_____

8.

_____

Problem Solving

9. Natasha has 1 quarter, 2 dimes, 10 nickels, and 4 pennies. She needs 1 dollar to buy a joke book. How much money does she have?

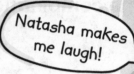

Natasha makes me laugh!

_____

How much more does she need to have one dollar?

_____

10. Chip needs 1 dollar. He has three quarters and one dime. How much does he have?

_____

How much more does he need to make 1 dollar?

_____

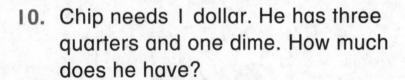

Write Math   Think of 2 combinations of coins that equal one dollar and write them here.

_____

_____

_____

Name _____

My Homework

Dollars

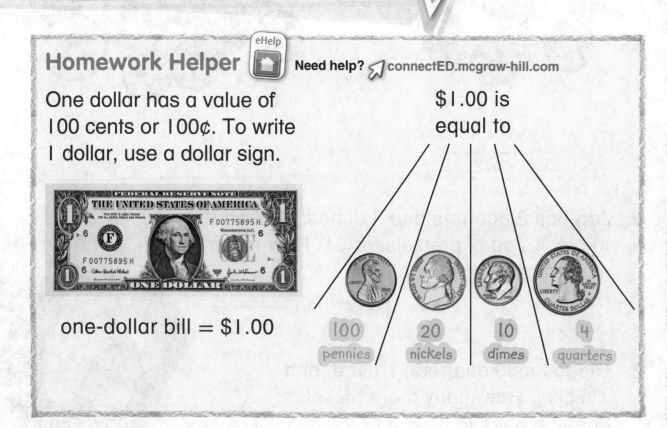

Homework Helper eHelp

Need help? connectED.mcgraw-hill.com

One dollar has a value of 100 cents or 100¢. To write 1 dollar, use a dollar sign.

one-dollar bill = $1.00

$1.00 is equal to

100 pennies   20 nickels   10 dimes   4 quarters

Count to find the value of the coins.
Circle combinations that equal $1.00.

1.

_____

2.

_____

Copyright © The McGraw-Hill Companies, Inc.   (tl)Michael Houghton/StudiOhio, (others)United States coin images from the United States Mint

Chapter 8 • Lesson 5   513

Count the coins. Write the value.
Circle combinations that equal $1.00.

3.

_____

4.

_____

5. Jen has 2 quarters and 4 dimes. She wants to buy 1 bag of pretzels for $1. How much more money does she need?

_____

6. Diego has 3 quarters, 1 dime, and 1 nickel. How many more nickels does he need to have $1?

_____ nickels

Vocabulary Check

Circle the correct choices.

7. **one dollar** $1   1$ $1.00   1¢

 Math at Home Have your child use various coins to show you two ways to make $1.

Name ....

My Review

Vocabulary Check

Draw lines to match.

1. **dime**   1 cent or 1¢

2. **penny**   25 cents or 25¢

3. **qu__ter**   5 cents or 5¢

4. **dollar**   10 ____ ___ or 10¢

5. **nickel**   100 cents or 100¢

Concept Check

Count to find the value of the coins.

6.

_____¢, _____¢, _____¢, _____¢, _____¢ = _____¢

Concept Check

Count to find the value of the coins.

7.

_____ ¢, _____ ¢, _____ ¢, _____ ¢, _____ ¢, _____ ¢ = _____ ¢

Count to find the value of the group of coins.

8.

_____ ¢, _____ ¢, _____ ¢, _____ ¢, _____ ¢, _____ ¢

= _____ ¢

Count to find the value of the coins.
Circle the combinations that equal $1.00.

9.

_____

10.

_____

Use each coin to make one dollar.
Write the number of coins you used.

11.

_____

12.

_____

Name _____

 Problem Solving

I only cost 47¢. I'm a bargain!

13. Lupe buys a toy dinosaur for 47¢. He gives the cashier 1 quarter and 1 dime. How much more money does he need to give the cashier?

_____

Circle the three coins Lupe should give the cashier.

14. John has two dimes. Mark has two quarters. How much money do the two boys have in all?

_____

Test Practice

15. Lacey found 1 quarter and 1 dime. She already had 30¢. Kyra has 85¢. How much more money does Kyra have?

10¢	20¢	25¢	30¢
○	○	○	○

I cent or 1¢

_____, _____, _____ = _____

_____ pennies make $1.00.

5 cents or 5¢

_____, _____, _____ = _____

_____ nickels make $1.00.

ESSENTIAL QUESTION
How do I count and use money?

10 cents or 10¢

_____, _____, _____ = _____

_____ dimes make $1.00.

25 cents or 25¢

_____, _____, _____ = _____

_____ quarters make $1.00.

Count on it!

Success is yours!

Chapter

9 Data Analysis

Our Bodies Need Healthful Food!

Watch a video!

Watch

My Standards

Measurement and Data

2.MD.9 Generate measurement data by measuring lengths of several objects to the nearest whole unit, or by making repeated measurements of the same object. Show the measurements by making a line plot, where the horizontal scale is marked off in whole-number units.

2.MD.10 Draw a picture graph and a bar graph (with single-unit scale) to represent a data set with up to four categories. Solve simple put-together, take-apart, and compare problems using information presented in a bar graph.

Standards for
Mathematical
PRACTICE

1. Make sense of problems and persevere in solving them.
2. Reason abstractly and quantitatively.
3. Construct viable arguments and critique the reasoning of others.
4. Model with mathematics.
5. Use appropriate tools strategically.
6. Attend to precision.
7. Look for and make use of structure.
8. Look for and express regularity in repeated reasoning.

= focused on in this chapter

Name

Am I Ready?

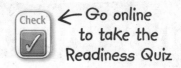 Check ✓ ← Go online to take the Readiness Quiz

Circle the group that has more.

1.

Use the tally chart to answer the questions.

2. What pet has 4 tally marks?

   _____

3. How many people like cats?

   _____ people

Favorite Pet	Tally
Cat	|||
Hamster	||||
Dog	⫼| ||

Use the picture to solve.

4. Roy walks four dogs every day.
   How many brown dogs does he walk?

   _____ brown dogs

How Did I Do?

Shade the boxes to show the problems you answered correctly.

1	2	3	4

Name

My Math Words

Vocab
abc

Review Vocabulary

compare   graph   tally marks

Use the review words to describe each example.

**What I Can Show
on a Graph?**

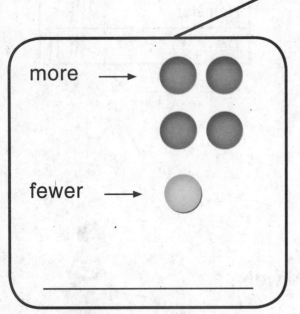

more →

fewer →

_____

How many balloons? ⫼⫽

_____

What is a graph?

_____

_____

_____

My Vocabulary Cards

Lesson 9–4

bar graph

Swimming Laps

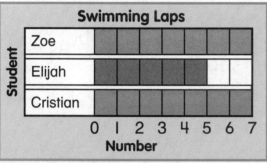

Student: Zoe, Elijah, Cristian

Number: 0 1 2 3 4 5 6 7

Lesson 9–1

data

What Pet Do You Like Best?	
Pet	Tally
Dogs	卌 卌 I
Cats	卌 II

Lesson 9–2

key

Favorite Vacation

Beach				
Camping				
Water park				

Key: Each picture = 1 vote

Lesson 9–7

line plot

Books Read in November

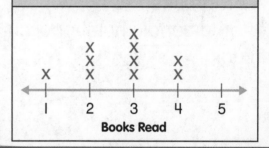

Books Read: 1 2 3 4 5

Lesson 9–2

picture graph

Our Favorite Toys

Balls							
Skates					
Stuffed animals				

Lesson 9–1

survey

Favorite Time of Day		Tally	Total
Morning		卌 I	6
Afternoon		III	3
Evening		卌 III	8

Teacher Directions:
Ideas for Use
• Have students group 2 or 3 common words. Ask them to add a word that is unrelated to the group. Have them work with a friend to name the unrelated word.

• Have students write a tally mark on each card every time they read the word in this chapter or use it in their writing.

Numbers or symbols that show information.

A graph that uses bars to show data.

A graph used to show how often a certain number occurs in data.

Tells what, or how many, each symbol stands for.

To collect data by asking people the same question.

A graph that has different pictures to show information collected.

Lesson 9–2

symbol

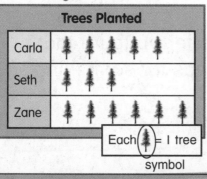

Trees Planted					
Carla	🌲	🌲	🌲	🌲	🌲
Seth	🌲	🌲	🌲		
Zane	🌲	🌲	🌲	🌲	🌲 🌲

Each 🌲 = 1 tree

symbol

Lesson 9–1

tally marks

Items Sold at School Store	
Item	Tally
Eraser	卌
Bottle of glue	卌 卌
Pencil	卌 III
Scissors	II

Teacher Directions:
More Ideas for Use

- Ask students to group like ideas they find throughout the chapter, such as using different ways to display data. Have students share the strategies they use to understand the concepts.

- Have students use a blank card to write this chapter's essential question. Have them use the back of the card to write or draw examples that help them answer the question.

Marks used in a survey to collect data.

A letter or figure that stands for something.

My Foldable

Name

....

Take a Survey

Lesson 1
ESSENTIAL QUESTION
How can I record
and analyze data?

Explore and Explain

So, what's your favorite fruit?

Oranges!

What is your favorite fruit?		
Fruit	Tally	Total
Apple		_____
Orange		_____
Grape		_____

Teacher Directions: Ask 10 classmates to name their favorite fruit.
Mark a tally for each choice. Write the total for each fruit. Which fruit
do most people like?

See and Show

When you take a **survey**, you ask a question. Use **tally marks** to record the answers, or data. **Data** is information.

I like scary books!

Taking a survey about books

1. Write a question.

2. Ask your question.

3. Record each person's answer with a tally mark.

4. Count the tally marks.

Type of Book	Tally	Total
Scary	||	2
Funny	⊮⊮ |||	8
Sports	⊮⊮	5

Ask 10 students their favorite season. Use tally marks to record the data.

Favorite Season	Tally	Total
Winter		
Spring		
Summer		
Fall		

Use the data in the chart to answer the questions.

1. What season do students like best?

   _____

2. How many students like winter and fall? Write a number sentence to solve.

   ____ ◯ ____ = ____

Talk Math How do tally marks help you organize data?

Name _____

On My Own

Ask 10 students their favorite picnic food. Use tally marks to record the data.

Helpful Hint

| means 1.

HHt means 5.

Picnic Food	Tally	Total
Hamburger		
Hot dog		
Chicken		

Use the data in the chart to answer the questions.

3. How many students like hamburgers? _____

4. How many students like hot dogs? _____

5. What is the favorite picnic food?

    _____

6. What is the least favorite picnic food?

    _____

7. How would the chart change if you added your

    favorite food? _____

    _____

Problem Solving

8. Claire took a survey of favorite snacks.
She marked 6 tallies for vegetables,
4 tallies for nuts, and 8 tallies for fruit.
How many friends did she survey?

_____ friends

9. Students took a survey for Fun Friday
activities. Seven students want a picnic.
Five students want to go to the zoo.
Eight students want to watch a movie.
Which chart shows these results?
Circle the correct chart.

Activity	Tally
Picnic	ЖЖ I
Zoo	ЖЖ
Movie	|||

How many students were surveyed?

Activity	Tally
Picnic	ЖЖ ||
Zoo	ЖЖ
Movie	ЖЖ |||

_____ students

HOT Problem How do you record data on
a tally chart?

_____

_____

Name ....

My Homework

Lesson 1

Take a Survey

Homework Helper

eHelp

Need help? connectED.mcgraw-hill.com

A tally chart shows the results of a survey.

Favorite Vegetable		Tally	Total
	Carrot	⳾⳾⳾⳾ꟾ	6
	Broccoli	ꟾꟾꟾ	3
	Corn	⳾⳾⳾⳾ ꟾꟾꟾ	8

Helpful Hint

| means 1.

⳾⳾⳾⳾ means 5.

Practice

Ask 10 people their favorite exercise. Use tally marks to record the data.

Use the data in the chart to answer the questions.

Favorite Exercise	Tally	Total
Jump Rope		
Run		
Dance		
Play Baseball		

1. How many people like to run and dance? _____

2. Do more people like to run and play baseball

    or dance and jump rope? _____

Ask 10 people their favorite snack. Use tally marks to record the data.

Use the data in the chart to answer the questions.

Favorite Snack	Tally	Total
Fruit		
Chips		
Cookies		
Carrot Sticks		

3. What snack do people like the most? _____

4. How many people like healthful

   snacks the most? _____

Ask 10 people their favorite juice flavor. Use tally marks to record the data.

Use the data in the chart to answer the questions.

Favorite Juice	Tally	Total
Apple		
Orange		
Cranberry		
Tomato		

5. What juice do people like the least? _____

6. How many people like cranberry juice

   and apple juice in all? _____

Wow! A lot of people like me!

Vocabulary Check

Circle the word that matches the definition.

7. Numbers or symbols that show information.

   tally marks   survey   data

 Math at Home Help your child to create a survey that they can give to family members.

Name ....

Make Picture Graphs

Explore and Explain

Lunchroom pals!

I love soup!

Favorite Lunch Food					
Sandwich					
Pizza					
Hot Dog					
Soup					

Key: Each food = 1 vote

Teacher Directions: Ask five classmates to pick their favorite lunch food.
Draw a picture in the chart to show each person's choice.

Online Content at ⌁ connectED.mcgraw-hill.com

See and Show

You show data with a **picture graph**.
The pictures are a **symbol** for the data.

Favorite Marble Color

Blue	◯	◯	◯		
Red	◯	◯	◯		

Key: Each marble = I vote

> **Helpful Hint**
> The key tells how many
> each symbol stands for.

Use the tally chart to make a picture graph.

I.

Favorite Shoes	Tally	Total
👟 Tennis Shoes	|||	3
🩴 Flip Flops	ⅢⅡ	5
👢 Boots	|	I

Favorite Shoes

Tennis Shoes					
Flip Flops					
Boots					

Key: Each shoe = I vote

Talk Math How are picture graphs different
from tally charts?

Copyright © The McGraw-Hill Companies, Inc.   Design Pics Inc./Alamy Images

Name ....

On My Own

Use each tally chart to make a picture graph.

2.

Favorite Drink	Tally	Total
Milk	|||	3
Apple Juice	||	2
Water	||||	4
Lemonade	||||	4

Favorite Drink

Milk					
Apple Juice					
Water					
Lemonade					

Key: Each drink = 1 vote

I'm not sticking out my tongue. That's just my straw!

3.

Favorite Sport	Tally	Total
Baseball	|||	3
Basketball	||||	4
Soccer	||	2
Football	||||	5

Favorite Sport

Baseball					
Basketball					
Soccer					
Football					

Key: Each ball = 1 vote

Problem Solving

Mathematical PRACTICE

Use the information to make a picture graph.

4. Lilian asked ten people their favorite flower. One said tulips. The same number of people said daisies and carnations. 5 said roses.

Favorite Flower

Tulip					
Daisy					
Rose					
Carnation					

Key: Each flower = I vote

5. A school asked 15 students their favorite subject. 3 students said science and 3 said reading. One less person said art than math.

Favorite Subject

Math					
Science					
Reading					
Art					

Key: Each book = I vote

Write Math Explain how picture graphs can be more helpful than tally charts.

_____

_____

_____

Name ....

Analyze Picture Graphs

Explore and Explain

I know I am!

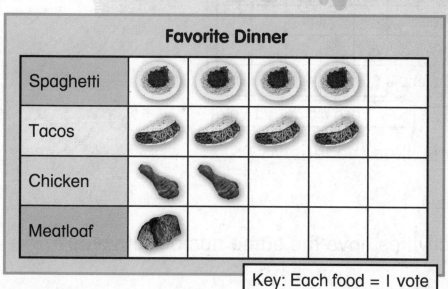

Favorite Dinner

Spaghetti					
Tacos					
Chicken					
Meatloaf					

Key: Each food = I vote

_____

_____

_____

_____

Which dinner is the least favorite? _____

Teacher Directions: Use the data in the picture graph. Write the number of votes for each dinner. Write the least favorite dinner.

See and Show

You can use a
picture graph to
answer questions.

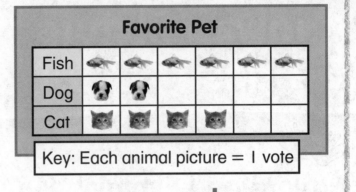

What pet is the favorite?
__fish__

How many votes does each picture show? _____

Use the data from the graph to answer the questions.

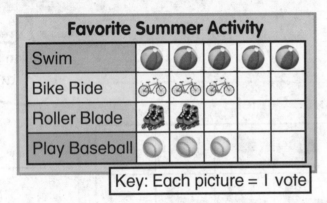

1. What activity is liked the least?

   _____

2. Which two activities have the same number of votes?

   _____

3. How many people like to swim and play baseball in all?

   _____

 How would you count the votes for bike
riding if each picture stood for 2 votes?

542   Chapter 9 • Lesson 3

Name _____

On My Own

Use the data from the graph to answer the questions.

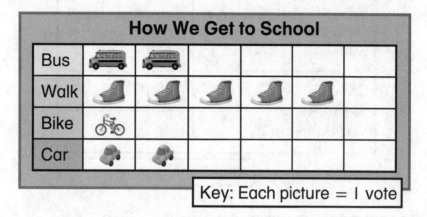

4. How do most students get to school? _____

5. How many students ride the bus and ride in cars? _____

6. How many students voted? _____

7. How will the graph change if you add another
   way to get to school?

   _____

8. Write a question that can be answered using the
   data in the picture graph.

   _____

   _____

9. Write a the number sentence you can use
   to answer your question.

   _____ ◯ _____ ◯ _____

Problem Solving

Use the data from the graph to answer the questions.

Favorite Pizza Toppings

Pepperoni	🍕	🍕	🍕	🍕	🍕
Sausage	🍕	🍕			
Cheese	🍕	🍕	🍕	🍕	
Vegetables	🍕	🍕	🍕		

Key: Each picture = 1 vote

10. Three more people were surveyed. They all like cheese. What topping is now the favorite?

_____

11. Carter voted for cheese. Kylie voted for vegetables. Whose favorite pizza topping got more votes?

_____

Write Math   Write a question about the graph above. Have a friend answer the question.

_____

_____

_____

Name

Measurement and Data
2.MD.10

My Homework

Lesson 3
Analyze Picture
Graphs

Homework Helper

Need help? connectED.mcgraw-hill.com

You can use the data from a picture graph to answer questions.

Favorite Vacation

Beach	🐚	🐚	🐚		
Camping	⛺	⛺			
Water park	🛟	🛟	🛟	🛟	

Key: Each picture = 1 vote

Which vacation got the most votes? __water park__

Practice

Use the data from the graph to answer the questions.

Favorite Hamburger Topping

Ketchup	🍾	🍾	🍾	🍾	
Mustard	🌭	🌭			
Lettuce	🥬				
Tomato	🍅				

Key: Each picture = 1 vote

1. How many people voted?

   _____

2. Which topping got the most votes? _____

3. Which two toppings got the least votes? _____

4. How many people like ketchup or tomato? _____

Chapter 9 • Lesson 3   545

Use the data from the graph to answer the questions.

Favorite Sandwich

Peanut butter and jelly	🥪	🥪	🥪		
Turkey	🥪	🥪			
Grilled cheese	🥪	🥪	🥪		
Peanut butter	🥪	🥪	🥪	🥪	

Key: Each sandwich = I vote

Vote for me! I'm for a longer lunch period!

5. How many people voted for turkey or grilled cheese? _____

6. Maria voted for peanut butter. She really wanted to vote for peanut butter and jelly. How would that change the picture graph?

_____

_____

_____

Test Practice

7. In the graph above, how many more people like grilled cheese than turkey?

I ⚪   2 ⚪   4 ⚪   7 ⚪

Math at Home Create a picture graph for your child about your family's favorite activities. Ask your child questions about the data on the graph.

Name _____

Check My Progress

Vocabulary Check

Write the word to complete each sentence.

picture graph   survey   key

1. In a _____, you collect data by asking people the same question.

2. A _____ tells what (or how many) each symbol stands for.

3. A _____ has different pictures to show information collected.

Concept Check

Ask 10 people their favorite color. Use tally marks to record the data.

Use the data in the chart to answer the questions.

Favorite Color	Tally	Total
Blue		
Pink		
Green		
Purple		

4. What color do people like the most? _____

5. How many people like pink and green? _____

6. What color do people like the least? _____

Use the tally chart to make a picture graph.

7.

Favorite Toy	Tally
Jump rope	|||
Baseball	||||
Roller blades	||
Bike	HH

Favorite Toy					
Jump rope					
Baseball					
Roller blades					
Bike					

Key: Each picture = 1 vote

Use the data from the graph to answer the questions.

8. How many more votes did bike get than baseball?

_____

9. How many people like jump ropes and roller blades?

_____

10. How many more people like to bike than jump rope?

_____

11. How many people like to jump rope and play baseball?

_____

Test Practice

12. Two students did not vote. If those 2 students all voted for roller blades, which toy would have the most votes?

Jump rope   Baseball   Roller blades   Bike

   ○   ○   ○   ○

Name ....

Make Bar Graphs

Explore and Explain ▶ Watch

I ski.

Favorite Winter Activity

Activity											
Sledding											
Ice Skating											
Skiing											
Snowboarding											

0 1 2 3 4 5 6 7 8 9 10
Number

Our class likes _____ the best.

Teacher Directions: Ask 10 classmates to pick their favorite winter activity. Color one square for each vote in the correct row. Write your classmates' favorite activity.

Online Content at ⌁ **connectED.mcgraw-hill.com**   Chapter 9 • Lesson 4   549

See and Show

Mathematical PRACTICE

A **bar graph** uses bars to show data. To make a bar graph, color one box for each vote. Bar graphs can look different.

Favorite Breakfast Food	Tally	Total
Toast	||||	4
Cereal	||||	5
Eggs	||	2

One Way

Another Way

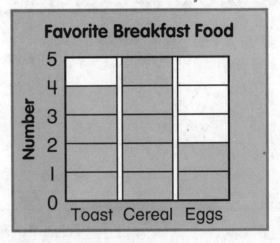

Use the tally chart to make a bar graph.

1.

Favorite Muffin	Tally	Total
Blueberry	||||	4
Strawberry	||	2
Lemon	|	1
Cherry	||||	5

Favorite Muffin

Muffin						
Blueberry						
Strawberry						
Lemon						
Cherry						

0 1 2 3 4 5
Number

Talk Math How are bar graphs different from picture graphs?

Copyright © The McGraw-Hill Companies, Inc.

550   Chapter 9 • Lesson 4

Name _____

On My Own

Complete the tally chart. Use the tally chart
to make a bar graph.

2.

Color of Shirt	Tally	Total
Red	IIII	
Blue	HHT I	
Black	II	
Green	HHT	

Color of Shirt

Colors	Red						
	Blue						
	Black						
	Green						

0 1 2 3 4 5 6
Number of Students

3.

Left-Handed or Right-Handed	Tally	Total
Left-Handed	IIII	
Right-Handed	HHT I	

Left-Handed or Right-Handed

Number of Students
7
6
5
4
3
2
1
0

Left-Handed   Right-Handed
Hands

Problem Solving

Use the data to fill in the bar graphs.

4. Alyssa took a survey to find which juice pop flavor 15 friends like.
6 voted for cherry.
5 voted for grape.
2 voted for orange. The rest of her friends voted for lime.

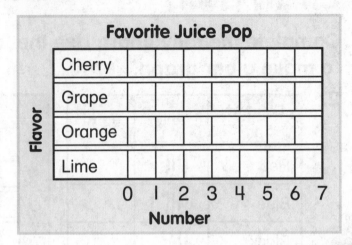

5. Luke took a survey of 18 friends to find their favorite soup. 4 voted for chicken soup. 6 voted for vegetable. The same number voted for tomato soup and noodle soup.

It's as easy as ABC!

HOT Problem Suppose each square in the Favorite Soup graph above was worth 2 votes. How would your bar graph change?

_____

_____

Name ....

Analyze Bar Graphs

Lesson 5

ESSENTIAL QUESTION
How can I record and analyze data?

Explore and Explain

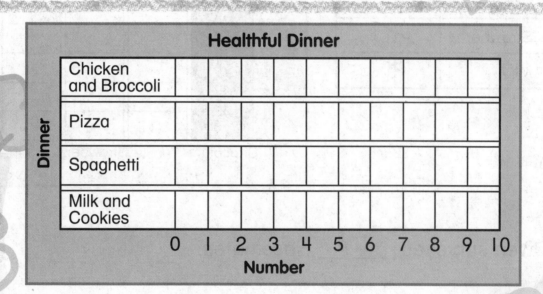

Healthful Dinner

Dinner											
Chicken and Broccoli											
Pizza											
Spaghetti											
Milk and Cookies											
	0	1	2	3	4	5	6	7	8	9	10

Number

chicken and broccoli _____

pizza _____

spaghetti _____

milk and cookies _____

Slurp!

Teacher Directions: Survey 10 people. Ask them which dinner is the healthiest. Make a bar graph. Write how many votes each dinner got.

See and Show

Owen surveyed his class about their favorite ice cream flavors. Then he made a bar graph to show the data.

Favorite Ice Cream Flavor	Tally
Chocolate	ⅢⅠ II
Vanilla	III
Strawberry	ⅢⅠ
Mint	ⅢⅠ I

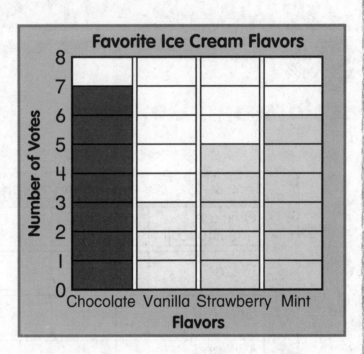

Owen surveyed ___21___ classmates.

Use the bar graph above to complete the sentences.

1. The flavor with the least votes is _____.

2. Chocolate received 2 more votes than _____.

3. What two flavors did eleven students vote for?

   _____.

 How could this graph help a teacher plan a class party?

Name _____

My Homework

Homework Helper Need help? connectED.mcgraw-hill.com

You can answer questions using the data from a bar graph.

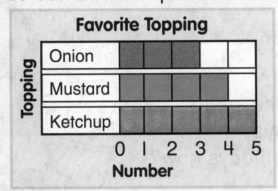

How many students were surveyed?

12 students were surveyed.

Practice

Use the bar graph to answer the questions.

1. How many people voted for mustard?

    _____

2. What topping do most people like?

    _____

3. Which topping is liked the least?

    _____

4. How many people like mustard or onion? _____

ketchup!

Use the bar graph to answer the questions.

5. How many people have April birthdays?

   _____

6. How many birthdays are in February and March?

   _____

7. How many people were surveyed in all?

   _____

8. Four more people answered the survey. They all had February birthdays. Which month has the most birthdays now?

   _____

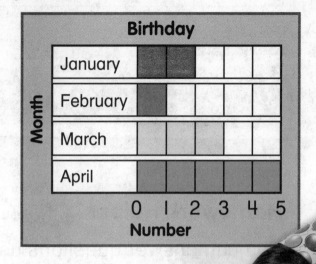

It's your big day!

Test Practice

9. How many birthdays are in May and August?

   8 ○   7 ○

   4 ○   3 ○

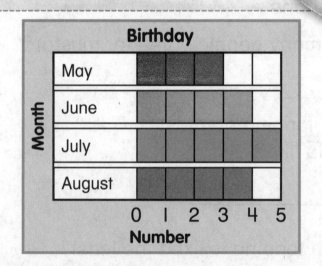

Math at Home Ask your child how he or she analyzed the bar graphs on this page.

Name

Problem Solving

STRATEGY: Make a Table

Lesson 6

ESSENTIAL QUESTION
How can I record and analyze data?

The breakfast special at Darla's Diner comes with 3 pancakes. Five friends order the breakfast special. How many pancakes will they get in all?

Good morning!

1 Understand Underline what you know.
Circle what you need to find.

2 Plan How will I solve the problem?

3 Solve Make a table.

___15___ pancakes

Friends	Pancakes
1	3
2	6
3	9
4	12
5	15

4 Check Is my answer reasonable? Explain.

Practice the Strategy

Each person in the class has the same number of pets. Two people have 4 pets. How many pets do 7 people have?

1 **Understand** Underline what you know.
Circle what you need to find.

2 **Plan** How will I solve the problem?

3 **Solve** I will...

Number of Pets	
People	Pets
1	2
2	4
3	
4	
5	
6	
7	

_____ pets

4 **Check** Is my answer reasonable? Explain.

Squawk!

You've got this one!

Squawk!

Chapter 9 • Lesson 6

Name _____

Apply the Strategy

1. Desiree has 4 pairs of socks in her drawer.
   How many socks are there in all?

Pair	Socks in All
1	2

We're clean! Really!

_____ socks

2. Mr. Minnick needs to deliver 60 boxes.
   His car can hold 10 boxes at a time.
   How many trips will he need to make
   to deliver all 60 boxes?

_____ trips

3. Juice boxes come in packs of 4 at Sam's
   Grocery. Mrs. Perez needs 20 juice boxes
   in all. How many packs should she buy?

_____ packs

Choose a strategy
- Make a table.
- Make a model.
- Find a pattern.

4. Aaron has 3 muffin pans. Each pan can hold 6 muffins. How many muffins can Aaron bake at a time?

_____ muffins

5. Suzie, Simon, and Sean each had 4 bottles of water during their softball game. How many bottles of water did they have in all?

_____ bottles

6. Mr. Bell wants his 21 students to put on their gloves to go outside. How many gloves do they have in all?

_____ gloves

7. Grace, Elijah, and Cameron are saving soup labels. Grace has 10, Elijah has 7, and Cameron has 13. How many more labels does Cameron have than Elijah?

_____ labels

My Homework

eHelp

Each hen lays the same number of eggs.
I hen lays 3 eggs. Two hens lay 6 eggs.
Three hens lay 9 eggs. How many eggs will
6 hens lay?

1 Understand Underline what you know.
Circle what you need to find.

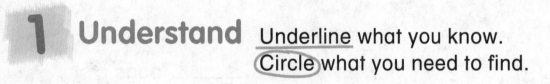

2 Plan How will I solve the problem?

3 Solve I will make a table.

Each hen lays 3 eggs.
So, 6 hens will lay
18 eggs.

Hens	Eggs
I	3
2	6
3	9
4	12
5	15
6	18

4 Check Is my answer reasonable?

Problem Solving

Underline what you know. Circle what you need to find.

1. Four turtles fit in one tank. Jose has 3 tanks. How many turtles can he have?

_____ turtles

2. Nine children want to feed the birds. They each have 2 bags of seed. How many bags of seed are there in all?

_____ bags

3. Shandra is giving a snack bag to each of her 4 friends. She puts 4 pear slices in each bag. How many pear slices are there in all?

_____ pear slices

We make a great pair!

4. Eight children are making snowmen. Each snowman needs 3 snowballs. How many snowballs do the children need in all?

_____ snowballs

Math at Home Have your child make a table to show how many meals he or she eats in a week.

Measurement and Data
Preparation for 2.MD.10

Make Line Plots

Lesson 7

ESSENTIAL QUESTION
How can I record and analyze data?

Explore and Explain

I'm ready to serve!

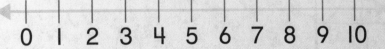

0 1 2 3 4 5 6 7 8 9 10

Teacher Directions: Chase asked 10 students how many servings of fruits and vegetables they eat each day. Three people eat 2 servings. Draw 3 X's over the number 2. Six people eat 5 servings. Draw 6 X's over the number 5. One person eats 7 servings. Draw 1 X over the number 7.

See and Show

A **line plot** is a way to organize data. Line plots are used to see how often a certain number occurs in data.

Mr. Sun's students marked their ages on a tally chart. They made a line plot using the data.

Age	Tally
7	|||
8	卌
9	||

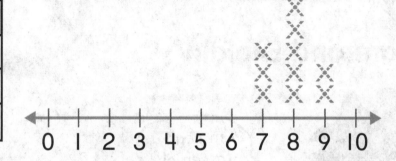

Use the tally chart to make a line plot.

1.

Pets	Tally
1	||
2	|||
3	||
4	|

0 1 2 3 4 5 6 7 8 9 10

2.

Sisters	Tally
0	||
1	|||
3	|
4	

0 1 2 3 4 5 6 7 8 9 10

Talk Math   How are line plots similar to tally charts?

Name _____

On My Own

Use the tally chart to make a line plot.

3.

Brothers	Tally
0				
1					
3				
4		

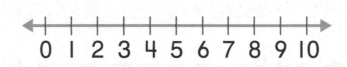

4.

Cousins	Tally
4				
5					
6	~~				~~	
7	~~||||~~

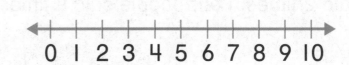

5.

Aunts	Tally
3	~~||||~~	
4				
5			
6				

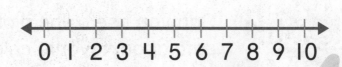

We are like one BIG family in Mr. Sun's class!

Problem Solving

Use the data to make a line plot.

6. Eli asked 10 classmates how many glasses of water they drink each day. Two people said 1. Three people said 0. Three people said 4. Everyone else said 6.

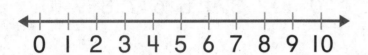

0 1 2 3 4 5 6 7 8 9 10

7. Sydney asked 15 friends how many times they exercise each week. The same number of people said 1 time and 3 times. Five people said 2 times. Four people said 4 times.

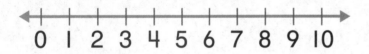

0 1 2 3 4 5 6 7 8 9 10

Write Math   Can you use a line plot to show data about favorite color?

_____

_____

Name ....

My Homework

Lesson 7

Make Line Plots

Homework Helper   Need help? connectED.mcgraw-hill.com

You can use data from a tally chart to make a line plot.
Line plots show how often a certain number occurs in data.

Sports	Tally
0			
1				
2	ﬀﬀﬀ
3				

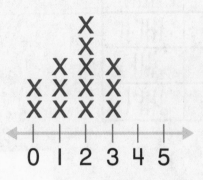

Practice

Use the tally chart to make a line plot.

1.

Siblings	Tally
0				
1		
2				
3			

Copyright © The McGraw-Hill Companies, Inc.   D. Hurst/Alamy Images

Use the tally chart to make a line plot.

2.

Second Grade Classes	Tally
1	I
2	III
3	HHT
4	HHT II

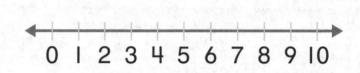

0 1 2 3 4 5 6 7 8 9 10

3.

Swings	Tally
5	IIII
6	HHT II
7	III
8	HHT I

0 1 2 3 4 5 6

Vocabulary Check

4. Circle the **line plot.**

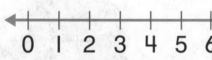

X
X
X X
X X X
X X X
0 1 2 3 4 5 6 7 8 9 10

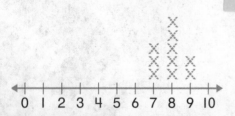

Our Favorite Toys

Balls		
Skates		
Stuffed animals		

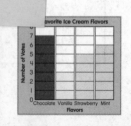

Favorite Ice Cream Flavors

Number of Votes

0 1 2 3 4 5 6 7

Chocolate Vanilla Strawberry Mint

Flavors

Math at Home Help your child take a survey of the ages of his or her cousins. Have your child make a line plot to show the data.

Name ....

Analyze Line Plots

Lesson 8

ESSENTIAL QUESTION
How can I record and
analyze data?

Explore and Explain

Should we split dessert?

0	1	2	3	4	5	6	7	8	9	10

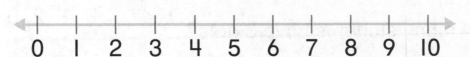

How many times a week do most people eat
dessert? _____

Teacher Directions: Ask 10 people how many times they eat dessert each week.
Use the data to make a line plot. Answer the question.

See and Show

You can use data from a line plot to answer questions.

This line plot shows the number of students in each class at a school.

Number of Students

```
              X
       X      X
       X      X
       X      X      X
  ←─┼──┼──┼──┼──┼──┼──┼──┼──┼──┼──┼──→
   15 16 17 18 19 20 21 22 23 24 25
```

How many students are in most of the classes?

Use the data from the line plot to answer the questions.

1. How many students have 2 hats?

   _____

2. How many students have just 1 hat?

   _____

3. How many students have 3 hats?

   _____

4. How many students have more than 2 hats?

   _____

Number of Hats

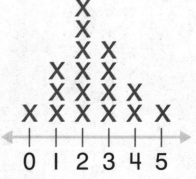

Talk Math How are line plots the same or different than bar graphs or picture graphs?

Name

On My Own

Use the data from the line plot to answer the questions.

Number of Vegetables

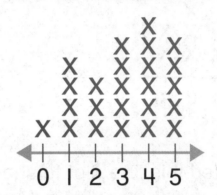

Carrot buzz cut

5. Which number has the most Xs? _____

6. How many students like 4 vegetables? _____

7. Which number has the fewest Xs? _____

8. How many students like 2 vegetables? _____

9. How many students were surveyed in all? _____

10. How many students like more than 1 vegetable?

    _____ students

11. How many students do not like vegetables?

    _____ student

Problem Solving

12. Finish the line plot for Ashley. She needs to show that three friends each have 2 toy bears.

Number of Toy Bears

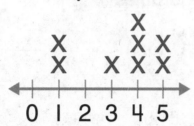

Teddy!

13. 23 students created this line plot to show how many fruits each student likes. How many students do not like fruit at all?

Fruits

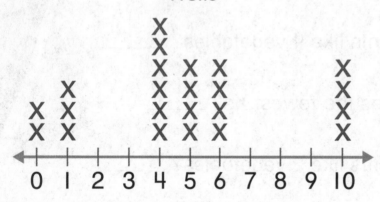

_____ students

Write Math Colton looks at the line plot above and says 1 student likes three fruits. Tell why he is wrong. Make it right.

_____

_____

_____

My Homework

Homework Helper Need help? connectED.mcgraw-hill.com

You can answer questions from data in line plots.

Number of Dolls

How many friends have 3 dolls?

5 friends each have 3 dolls.

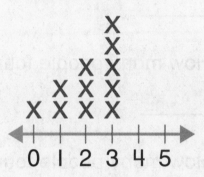

Practice

Use the data from the line plot to answer the questions.

1. How many people get $5? _____

2. How much allowance do most people get?

   _____

3. How many people get more than $6?

   _____

My Friends' Allowances

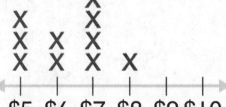

Use the data from the line plot to answer the questions.

Number of Pennies

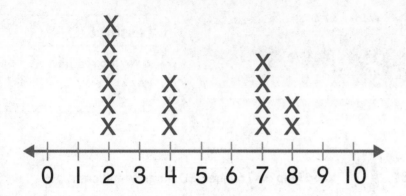

4. How many people found pennies?

_____

5. How many people found more than 5 pennies?

_____

6. How many people found less than 5 pennies?

_____

Test Practice

7. How many people have 5 dogs?

3   2 ○

1 ○   0 ○

Number of Dogs

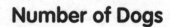
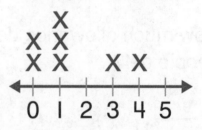

Math at Home Make a line plot about the number of healthy food choices your child makes each day. Ask your child questions about the line plot at the end of a week.

Name _____

My Review

Vocabulary Check

| data | survey | tally marks |
| picture graph | line plot | bar graph |

Write the correct word in each blank.

1. You can collect data by asking people the same question in a _____.

2. A _____ uses bars to show data.

3. Numbers or symbols that show information are called _____.

4. A _____ shows how often a certain number occurs in data.

5. A _____ uses pictures to show information collected.

6. A _____ is a mark used to record data collected in a survey.

Concept Check ✓

Use the tally chart to make a picture graph.

7.

Favorite Sport	
Sport	Tally
Baseball	IIII
Football	IIII
Hockey	II
Volleyball	III

Favorite Sport

Baseball					
Football					
Hockey					
Volleyball					

Key: Each picture = 1 vote

Use the picture graph above to complete the sentences.

8. How many students were surveyed? _____

9. How many students like hockey and football? _____

Use the tally chart to make a bar graph.

10.

Have you been to the zoo?	Tally
Yes	HHT HHT
No	I

Have you been to the zoo?

Yes										
No										

0 1 2 3 4 5 6 7 8 9 10
Number of Students

Use the bar graph above to complete the sentences.

11. How many students have been to the zoo? _____

12. How many more students have been to the zoo than have not been to the zoo? _____

Name _____

Problem Solving

13. Taylor asked 12 friends to name their favorite number. Five friends said 4. Six friends said 1. One friend said 2. Use Taylor's data to complete the line plot.

My word problem isn't way up here. It's at the bottom of the page.

Our Favorite Numbers

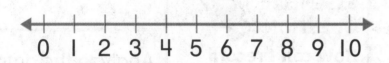

0 1 2 3 4 5 6 7 8 9 10

14. A tricycle has 3 wheels. How many wheels are there on 4 tricycles?

_____ wheels

Test Practice

15. There are 4 giraffes at the zoo. Each giraffe has 4 legs. How many legs are there in all?

2 ○   4 ○   8 ○   16 ○

Reflect

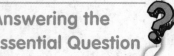

Show how to record and analyze data.

Complete the tally chart.

Fruit	Tally	Total
Banana	卌	
Apple	||||	
Orange	|||	

Use the tally chart to make a picture graph.

Fruit					
Banana					
Apple					
Orange					

Key: Each picture = 1 vote

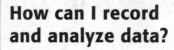

ESSENTIAL QUESTION

How can I record and analyze data?

Use the tally chart to make a bar graph.

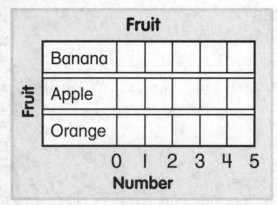

Fruit

Banana					
Apple					
Orange					

0 1 2 3 4 5
Number

Analyze the data.

How many people like apples and bananas?

_____

You can do it!

Keep your eyes on your goal!

ESSENTIAL QUESTION
How do I use and tell time?

Watch a video!

My Standards

Measurement and Data

2.MD.7 Tell and write time from analog and digital clocks to the nearest five minutes, using A.M. and P.M.

Standards for
Mathematical
PRACTICE

1. Make sense of problems and persevere in solving them.
2. Reason abstractly and quantitatively.
3. Construct viable arguments and critique the reasoning of others.
4. Model with mathematics.
5. Use appropriate tools strategically.
6. Attend to precision.
7. Look for and make use of structure.
8. Look for and express regularity in repeated reasoning.

= focused on in this chapter

Name

Am I Ready?

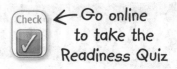 ← Go online to take the Readiness Quiz

Circle the time.

1.    7 o'clock   2 o'clock   6 o'clock

2.    4:00   5:00   6:00

Find the missing numbers.

3.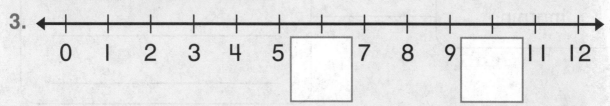

0   1   2   3   4   5   ☐   7   8   9   ☐   11   12

4. 2, 4, _____, 8, _____, 12, 14, 16, _____, 20

5. 5, _____, _____, 20, _____, _____, 35, 40, _____

6. Kira is walking home from school. Circle the time of day for the activity.

   morning   afternoon   evening

How Did I Do?

Shade the boxes to show the problems you answered correctly.

1	2	3	4	5	6

Name

My Math Words

Vocab

Review Vocabulary

afternoon   evening   morning

Write about an activity you do for each time of day.
Use *morning*, *afternoon*, or *evening* in each sentence.

morning

afternoon

evening

My Vocabulary Cards

Vocab
abc

Mathematical
PRACTICE

Lesson 10–6

A.M.

Lesson 10–1

analog clock

Lesson 10–1

digital clock

Lesson 10–2

half hour

half past 5
5:30

Lesson 10–1

hour

hour
3 o'clock
3:00

Lesson 10–1

hour hand

Teacher Directions:
Ideas for Use
Directions:

• Tell students to create riddles for each word. Ask them to work with a friend to guess the word for each riddle.

• Ask students to arrange cards to show an opposite pair. Have them explain the meaning of their pairing.

A clock that has an hour hand and a minute hand.

The hours from midnight until noon.

One half hour is 30 minutes. Sometimes called half past.

A clock that uses only numbers to show time.

The shorter hand on a clock that tells the hour.

A unit to measure time.
I hour = 60 minutes

My Vocabulary Cards

Mathematical
PRACTICE

Lesson 10-2

minute

I minute
60 seconds

Lesson 10-1

minute hand

Lesson 10-6

P.M.

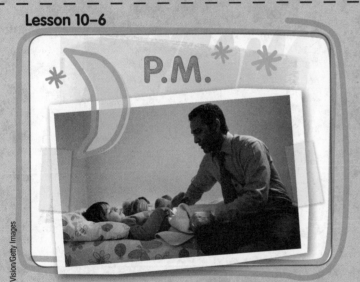

Lesson 10-4

quarter hour

quarter past 12   quarter till 1

The longer hand on a clock that tells the number of minutes.

A unit to measure time. Each tick mark on an analog clock.
1 minute = 60 seconds
60 minutes = 1 hour

A quarter hour is 15 minutes. Sometimes called quarter past or quarter till.
4 quarter hours = 1 hour

The hours from noon until midnight.

My Foldable

FOLDABLES® Follow the steps on the back to make your Foldable.

✂ — — — — — — — — — — — — — — — — — —

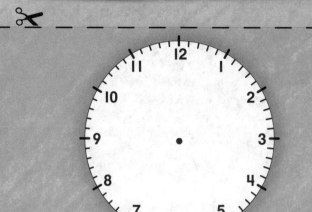

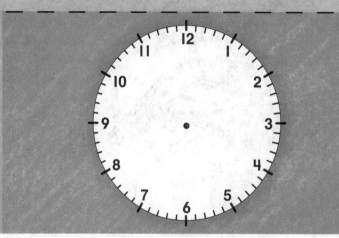

A.M.   P.M.

A.M.   P.M.

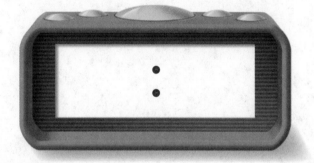

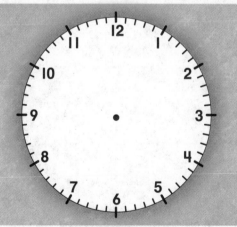

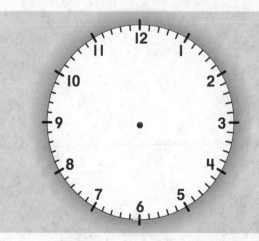

A.M.   P.M.

A.M.   P.M.

1 o'clock

quarter till 9

quarter past 2

half past 3

Name
....

Time to the Hour

Explore and Explain

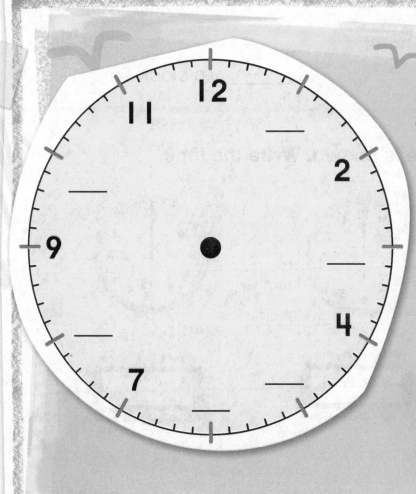

This is a
cool way to get
to school!

I hope
we're not
late.

_____ o'clock

 Teacher Directions: Write the missing numbers on the clock.

Use 🕐 to show time to each hour. Draw hands on the clock to show
4 o'clock. Write the time.

Online Content at ⚡ **connectED.mcgraw-hill.com**

See and Show

On an **analog clock**, the **hour hand** points to the **hour**. It is shorter.

A **digital clock** shows the hour and minutes on a screen.

The **minute hand** points to the **minute**. It is longer.

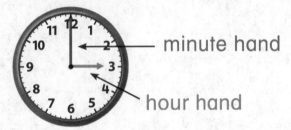

minute hand

hour hand

hour minutes after the hour

_3_ o'clock

Use . Tell what time is shown. Write the time.

1.

2.

3.

_____ o'clock

_____ o'clock

_____ o'clock

Talk Math How is an analog clock similar to a number line?

Name

On My Own

Use . Tell what time is shown. Write the time.

4.

_____ o'clock

5.

_____ o'clock

6.

_____ o'clock

Draw the hands on the clock. Write the time.

7. 5 o'clock

8. 10 o'clock

9. 3 o'clock

Problem Solving

Show the time on each clock.

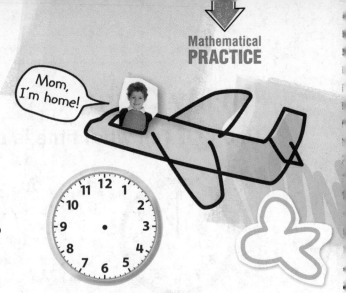

Mom, I'm home!

10. Colin gets home at 3 o'clock. Evan gets home one hour later. What time does Evan get home?

11. Basketball practice starts at 4:00. Mark eats a snack 2 hours before practice. What time does Mark eat his snack?

12. It is after 7 o'clock and before 9 o'clock. What time is it if it is exactly on the hour? Show the time on the analog clock face.

Write Math What time is it when the hour hand is on 5 and the minute hand is on 12? How do you know?

_____

_____

_____

My Homework

Homework Helper

eHelp

Need help? connectED.mcgraw-hill.com

You can tell and write time to the hour.

minute hand

hour hand

It is 5 o'clock or 5:00.

hour   minutes after the hour

Practice

Tell what time is shown. Write the time.

1.

_____ o'clock

2.

_____ o'clock

3.

_____ o'clock

Draw the hands on each clock. Then write the time.

4. 2 o'clock

5. 4:00

6. The time on Javier's watch is 3 hours after 2:00. Write the time on his watch.

Javier's cool watch!

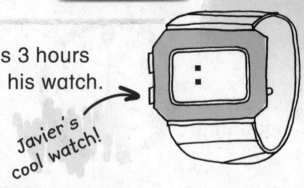

Vocabulary Check

Complete each sentence.

hour hand   analog clock   minute hand   digital clock

7. An _____ has an hour and a minute hand.

8. The _____ points to the hour on an analog clock.

9. A _____ uses numbers on a screen to show time.

10. The _____ points to the minute on an analog clock.

Math at Home Throughout the day, ask your child to look at an analog or digital clock showing a time to the hour and have them tell you the time.

Name ....

Time to the Half Hour

Lesson 2
ESSENTIAL QUESTION
How do I use and tell time?

Explore and Explain

It's time for our trip!

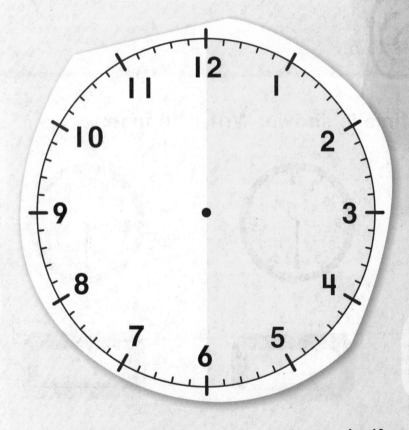

half past _____

Teacher Directions: Use a to show 5 o'clock. Move the minute hand to 6. What time is it? Repeat with other times showing hour and half past the hour. Draw hands to show one of the times and write the time.

Online Content at **connectED.mcgraw-hill.com**   Chapter 10 • Lesson 2   599

See and Show

You can show time to the half hour.
One half hour is 30 minutes.
These clocks show **half past** 2.

Helpful Hint
The hour hand moves halfway to the next number as a half hour passes.

The hour hand points between 2 and 3. The minute hand points to 6.

It is 2:30.

Use . Tell what time is shown. Write the time.

1.

half past _____

2.

half past _____

3.

half past _____

Talk Math It is half past 8. Explain what *half past* means.

Name

On My Own

Use . Tell what time is shown. Write the time.

4.

half past _____

5.

half past _____

6.

half past _____

Draw the hands on each clock. Write the time.

7. half past 4

8. half past 7

9. half past 6

Problem Solving

10. Eli woke up at half past 7. He needs to leave for school one hour later. Show and write the time he needs to leave for school.

Use the clocks shown below.

School   Soccer   Alarm
starts   practice   goes off

What happens at each of these times?

11. 6:30 _____

12. 3:30 _____

13. 8:30 _____

 How is reading a digital clock different from reading an analog clock? Explain.

_____

_____

Name _____

My Homework

Homework Helper Need help? connectED.mcgraw-hill.com

You can tell and write time to the half hour.

hour hand

minute hand

hour minutes after hour

It is half past 3 or 3:30.

Practice

Tell what time is shown. Write the time.

1.

half past _____

2.

half past _____

3.

half past _____

Draw the hands on each clock. Write the time.

4. half past 1

5. half past 11

6. Tatum is going to the city with her family. They will leave one hour after she gets home from school. Tatum gets home from school at 3:30. What will the clocks look like when they leave?

No, we're not lost. We're on an adventure!

Vocabulary Check

Show a **half hour** later on the second clock.

7.

Math at Home Give your child a time to the hour. Have him or her tell you the position of the clock hands for half past that hour or in a half hour.

Problem Solving

STRATEGY: Find a Pattern

Measurement and Data
2.MD.8

Lesson 3

ESSENTIAL QUESTION
How do I use and tell time?

The buses leave in order every half hour. Bus 1 leaves at 9:30. Bus 2 leaves at 10:00. What time do buses 3 and 4 leave?

All aboard!

1 Understand
Underline what you know.
Circle what you need to find.

2 Plan
How will I solve the problem?

3 Solve
Find a pattern.

9:30 , 10:00 , 10:30 , 11:00

Bus 1 Bus 2 Bus 3 Bus 4

4 Check
Is my answer reasonable?

Practice the Strategy

Each class at school will go to recess 30 minutes after the other. The first class goes to recess at 1:00. What time will the fourth class go?

I'm going to the moon!

Blast off!

1 Understand Underline what you know.
Circle what you need to find.

_____

2 Plan How will I solve the problem?

_____

3 Solve I will...

_____, _____, _____, _____

  Class 1   Class 2   Class 3   Class 4

_____

4 Check Is my answer reasonable? Explain.

Apply the Strategy

Find a pattern to solve. Use the space provided.

1. In the morning, Ms. White's students change activities every hour. Reading starts at 8:30. Her students go to three more activities. When do each of the activities start?

    8:30, _____, _____, _____

2. Students start work in learning stations at 10:30. After two hours, they go to recess. What time do students go to recess?

    _____

3. In the afternoon, Ms. White's students change activities every half hour. Math starts at 1:00. Writing starts after math. What time will writing start?

    _____

Review the Strategies

Choose a strategy
- Find a pattern.
- Act it out.
- Draw a picture.

4. Carmen has 8 shoes. She places an equal number of shoes in two suitcases. How many shoes will be in each suitcase?

_____ shoes

5. A postcard costs 45¢. Jason wants to buy 2 postcards. He has 2 quarters, 2 dimes, and 3 nickels. How much money does he have?

_____

How many pennies does he need to so that he can buy the postcards?

_____ pennies

6. The subway arrives at the station on the hour and half hour. It is now 4:30. Kate's family is on the way but it will take one hour to get there. What is the earliest time her family can get on the subway?

_____

Name ....

My Homework

Lesson 3

Problem-Solving:
Find a Pattern

Ashley's family is going to visit their grandma. It will take them four hours to get there. They leave at 10:30. What time will they get to grandma's house?

1 Understand
Underline what you know.
Circle what you need to find.

2 Plan
How will I solve the problem?

3 Solve
Find a pattern.

10:30   11:30   12:30   1:30   2:30

I hour   I hour   I hour   I hour

They will arrive at 2:30.

4 Check
Is my answer reasonable?

Problem Solving

Underline what you know. Circle what you need to find. Find a pattern to solve.

1. There is one plane that leaves from the city to the ski resort every hour. The first plane leaves at 10:00. The last plane leaves at 4:00. How many planes fly to the resort each day?

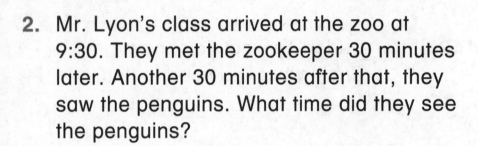

_____ planes

2. Mr. Lyon's class arrived at the zoo at 9:30. They met the zookeeper 30 minutes later. Another 30 minutes after that, they saw the penguins. What time did they see the penguins?

_____

3. Julio's family wants to take a ride on a tour bus. It leaves the bus station every half hour. It is 1:00. They just missed the bus. What are the times for the next two buses?

_____ , _____

Math at Home Ask your child to keep a journal for one evening. At each half hour, have your child write the time and record his or her activity at that time. At the end of the evening, see if there were any repeating or other patterns in their activities.

Name _____

Check My Progress

Vocabulary Check

Complete each sentence.

minute   analog clock   hour

digital clock   half hour

1. 60 minutes is one _____.

2. 30 minutes is one _____.

3. 60 seconds is one _____.

4. A _____ uses only numbers to show time.

5. An _____ has an hour hand and minute hand to show time.

Concept Check

Tell what time is shown. Write the time.

6.

7.

8.

Draw the hands on each clock. Write the time.

9. half past 8

10. 11 o'clock

11. half past 1

Write the time.

12.

half past _____

13.

_____ o'clock

14. Kevin has three classes in a row. Each class is 2 hours long. Kevin's first class is at 7:00. When will Kevin's last class end?

_____

Test Practice

15. The class bell rings every 30 minutes. The first one rings at 8:30. What time will the 4th bell ring?

10:30   10:00   11:30   9:30
  ○   ○   ○   ○

Name ....

Time to the Quarter Hour

Lesson 4

ESSENTIAL QUESTION
How do I use and tell time?

Explore and Explain

I can lend a hand with quarter hours!

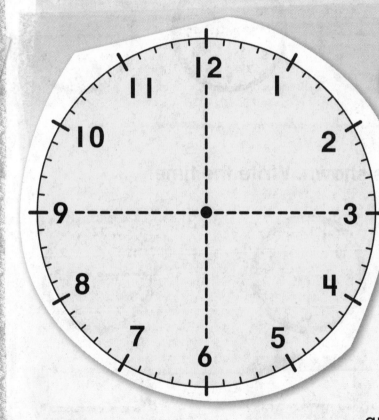

quarter past _____

Teacher Directions: Have students fold a paper plate, or circular piece of paper in half. Fold it again in quarters. Unfold and look at the fold marks. On the clock above, place cubes on the quarter hours referring back to their plate or circle. Draw hands on the clock to show quarter past an hour. Write the time.

See and Show

You can show time to the **quarter hour**. There are
15 minutes in a quarter hour. There are 4 quarter hours
in an hour.

1 o'clock	quarter past 1	half past 1	quarter till 2
1:00	1:15	1:30	1:45

Use 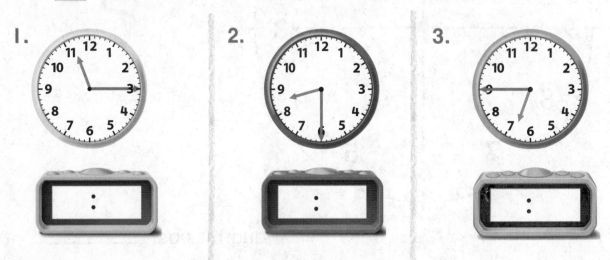. Tell what time is shown. Write the time.

1.

2.

3.

Talk Math   At 4:15, where is the minute
hand? Explain.

Name

On My Own

Use . Tell what time is shown. Write the time.

4.

5.

6.

Draw the hands on each clock. Write the time.

7. quarter till 2

8. quarter past 5

9. quarter past 12

Read the time. Write the time on the digital clock.

10. twelve forty-five

11. four fifteen

12. nine thirty

Problem Solving

Mathematical
PRACTICE

13. Mia is at school. School is over at 3:15. She has 3 more hours to wait. What time is it now?

_____

14. Ken goes to his friend's house at 2:15. Circle the clock that shows this time.

15. Alea's family went hiking. They left at 9:45. They drove for 3 hours. Then they stopped for lunch. Lunch took one hour. What time did they finish lunch?

Follow me!

_____

HOT Problem Why is each 15-minute period on a clock called one quarter hour?

_____

_____

_____

Copyright © The McGraw-Hill Companies, Inc.   Jupiter Images/BananaStock/Alamy Images

My Homework

Lesson 4

Time to the
Quarter Hour

Homework Helper Need help? 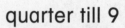connectED.mcgraw-hill.com

You can show time to the quarter hour.

quarter past 8   half past 8   quarter till 9

Practice

Use . Tell what time is shown. Write the time.

I.

2.

3.

Draw the hands on each clock. Write the time.

4. quarter past 6

5. quarter till 3

6. quarter past 7

7. Carly left for school at 8:15. She left school at 2:45. If she arrived at school at 8:45, how many hours was she in school?

The day flew by!

_____ hours

Vocabulary Check

8. Circle the clock that shows an example of quarter hour.

7:35

Copyright © The McGraw-Hill Companies, Inc.
Mike Kemp/RubberBall Productions/Getty Images

Math at Home Have your child use the words quarter till and quarter past to describe the time at 6:15 and 6:45.

Measurement and Data
2.MD.7

Time to Five Minute Intervals

Lesson 5

ESSENTIAL QUESTION
How do I use and tell time?

Explore and Explain **Tools**

Please fasten your seat belts. This plane leaves in 5 minutes!

 Teacher Directions: Count by 5s as you trace the dotted line. Label each jump 5, 10, 15 and so on. Draw the minute hand. Write that time on the digital clock.

Online Content at connectED.mcgraw-hill.com

See and Show

It takes 5 minutes for the minute hand to move to the next number. You can skip count by 5s to tell the time.

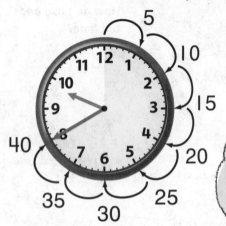

This clock shows ___40___ minutes after 9 o'clock. Write the time another way.

Helpful Hint
Each mark on the clock face is one minute.

Tell what time is shown. Use 🕐 to help. Write the time.

1.

2.

3.

4. Tell what time is shown. Draw the minute hand to show the time.

Talk Math Explain how you skip count by 5s to tell time.

Name

On My Own

Tell what time is shown. Use 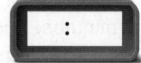 to help. Write the time.

5.

6.

7.

Tell what time is shown. Draw the hands to show the time.

8.

9.

10.

11. 2:20

12. 8:45

13. 1:25

Problem Solving

Mathematical
PRACTICE

14. If the hour hand is close to the 11 and the minute hand is pointing to the 10. What time is it?

_____

I can see my house from here!

15. A group of people get on an amusement park ride every 5 minutes. It is now 3:00. There are 7 groups ahead of Alyssa and her family. What time will her family go on the ride?

16. It takes 20 minutes to watch the movie previews. The previews start at 7 o'clock. What time will the movie start?

Write Math What time is it when the hour hand is between 5 and 6 and the minute hand is on 7? Explain.

_____

_____

_____

My Homework

Lesson 5

Time to Five-
Minute Intervals

Homework Helper

Need help? connectED.mcgraw-hill.com

It takes 5 minutes for the minute hand to move to the next number. Skip count by 5s to tell time.

This clock shows 35 minutes after 4 o'clock.

The time is shown another way.

Practice

Read the time. Write the time.

1.

2.

3.

Tell what time is shown. Draw the minute hand to show the time.

4.

5.

6.

7. It is 1:00. Hunter is waiting for Ben. Ben said he would meet Hunter in 25 minutes but he arrived 5 minutes late. What time does Ben meet Hunter?

Test Practice

8. A train leaves at 10:40. Which clock shows 10:40?

 Math at Home Ask your child to look at an analog clock on the hour. Have him or her tell you what time it will be in 5 minutes, 10 minutes, 25 minutes, and 50 minutes.

Copyright © The McGraw-Hill Companies, Inc.   (t)Steve Allen Travel Photography/Alamy Images, (b)Image Source/Punchstock

Name ....

A.M. and P.M.

Lesson 6

ESSENTIAL QUESTION
How do I use and
tell time?

Explore and Explain

Whoo hoo!

Where do you dream of traveling to?

Teacher Directions: Think of things that you do during the day and at night.
Draw pictures of the activities on the correct side of the page.

See and Show

Mathematical PRACTICE

The hours from midnight until noon are labeled **A.M.** The hours from noon until midnight are labeled **P.M.**

Helpful Hint
12:00 P.M. is noon.
12:00 A.M. is midnight.

Go to School

8:00 A.M.

Read a Bedtime Story

8:00 P.M.

Tell what time is shown for the activity. Write the time. Circle A.M. or P.M.

I. Art class

 A.M.
P.M.

2. Go to Bed

 A.M.
P.M.

3. Play after School

 A.M.
P.M.

Talk Math   How can you remember if it is A.M. or P.M.?

Name

On My Own

**Tell what time is shown for the activity. Write the time.
Circle A.M. or P.M.**

4. Eat Breakfast

 A.M.
P.M.

5. Wash the Dog

 A.M.
P.M.

6. Go Swimming

 A.M.
P.M.

**Tell what time is shown for the activity. Draw the
hands on the clock to show the time. Circle A.M. or P.M.**

7. Soccer Practice

 A.M.
P.M.

8. Sleeping in Bed

 A.M.
P.M.

9. Eat Dinner

 A.M.
P.M.

Complete each sentence. Write *midnight* or *noon*.

midnight   noon

10. At _____, I am usually asleep.

11. At _____, I may be eating my lunch.

Problem Solving

12. Zack has lacrosse practice at 4:00. Would that likely be at 4:00 A.M. or 4:00 P.M.?

_____

13. It is 2:25. Lakota is going to the library in 1 hour. On the clock show the time she will go to the library.
Circle A.M. or P.M.

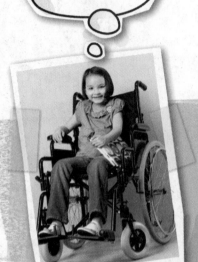

Library time is my favorite time of day!

A.M.   P.M.

14. It is 3:00 A.M. Are you more likely to be on your way home from school or sleeping in your bed?

_____

Write Math   Christy wants to go to her friend Jill's party. The invitation says the party starts at 1:30 A.M. This is wrong. What mistake was made on the invitation?

_____

_____

_____

Name _____

My Homework

Lesson 6

A.M. and P.M.

Homework Helper

eHelp

Need help? connectED.mcgraw-hill.com

The hours from midnight until noon are labeled A.M. The hours from noon until midnight are labeled P.M.

Helpful Hint
12:00 P.M. is noon.
12:00 A.M. is midnight.

Wake Up

9:00 A.M.

Look at the Moon

9:00 P.M.

Tell what time is shown for the activity. Write the time. Circle A.M or P.M.

1. Go to the Park

[:] A.M.
P.M.

2. Go Bowling

[:] A.M.
P.M.

3. Make your Bed

[:] A.M.
P.M.

Tell what time is shown for the activity.
Draw the hands on the clock. Circle A.M. or P.M.

4. Rooster Crowing

 A.M.
P.M.

5. Flying a Kite

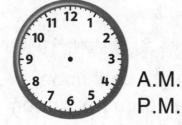 A.M.
P.M.

6. Going Shopping

 A.M.
P.M.

7. Jay and Mike went to the county fair with Mike's parents. They got there at 10:00 A.M. They stayed for 4 and a half hours. Did they get home in the A.M or P.M.? _____

Vocabulary Check

Complete each sentence.

A.M.   P.M.

8. Casey has an art lesson at 4:30. _____.

9. Sam eats dinner at 5:30 _____.

10. Kaylee eats breakfast at 6:00 _____.

 Math at Home Several times in the next 24 hours, ask your child for the time and then if it is A.M. or P.M.

Name _____

Vocabulary Check

Complete each sentence.

hour   half hour   quarter hour   minute
A.M.   P.M.   digital   analog

1. The long hand on an analog clock is the _____ hand.

2. The short hand on an analog clock is the _____ hand.

3. 15 minutes is a _____.

4. 30 minutes is a _____.

5. A clock that uses hands to show the time is an _____ clock.

6. A clock that uses numbers to show the time is a _____ clock.

7. The hours from noon to midnight are _____.

8. The hours from midnight to noon are _____.

Concept Check

Read the time. Write the time.

9.

10.

11.

Tell what time is shown. Draw the hands on each clock.

12.

13.

14.

Circle the best choice.

15. Walk the dog.   11:45   A.M.   P.M.

16. Go hiking.   2:30   A.M.   P.M.

17. Wash dinner dishes.   6:30   A.M.   P.M.

Name ....

Problem Solving

18. On a clock, the hour hand is between the 2 and 3. The minute hand is on the 5. What time is it? Show the time on both clocks.

---

19. The time is 9:00. In 6 hours, Greg will leave school. What time will it be in 6 hours? Circle A.M. or P.M.

_____ A.M.   P.M.

---

20. Derek has baseball practice at 2:30. Luke has practice at 4:15. Show the times on the clocks. Write the time and circle if it would be A.M. or P.M.

Derek   Luke

   A.M.      A.M.
_____   P.M.   _____   P.M.

Test Practice

21. The boat picked up passengers every half hour. If the first boat arrives at 8:00 A.M., when will the fourth boat arrive?

12:00 P.M.   10:00 A.M.   9:30 A.M.   10:30 A.M.
   ○   ○   ○   ○

Reflect

Chapter 10

Answering the Essential Question

Show the ways to write and use time.

Draw the hands to show the time.

 Digital

 Analog

Write the time.

ESSENTIAL QUESTION

How do I use and tell time?

Skip count by 5s. Tell the time.

Skip count by fives.

The time is _____ .

Draw the hands. Write the time.

quarter past 5   half past 5

You'll sail through this!

Chapter 11

Customary and Metric Lengths

ESSENTIAL QUESTION
How can I measure objects?

I Love Sports!

Watch a video!

Watch

My Standards

Measurement and Data

2.MD.1 Measure the length of an object by selecting and using appropriate tools such as rulers, yardsticks, meter sticks, and measuring tapes.

2.MD.2 Measure the length of an object twice, using length units of different lengths for the two measurements; describe how the two measurements relate to the size of the unit chosen.

2.MD.3 Estimate lengths using units of inches, feet, centimeters, and meters.

2.MD.4 Measure to determine how much longer one object is than another, expressing the length difference in terms of a standard length unit.

2.MD.5 Use addition and subtraction within 100 to solve word problems involving lengths that are given in the same units, e.g., by using drawings (such as drawings of rulers) and equations with a symbol for the unknown number to represent the problem.

2.MD.6 Represent whole numbers as lengths from 0 on a number line diagram with equally spaced points corresponding to the numbers 0, 1, 2, …, and represent whole-number sums and differences within 100 on a number line diagram.

2.MD.9 Generate measurement data by measuring lengths of several objects to the nearest whole unit, or by making repeated measurements of the same object. Show the measurements by making a line plot, where the horizontal scale is marked off in whole-number units.

Standards for
Mathematical
PRACTICE

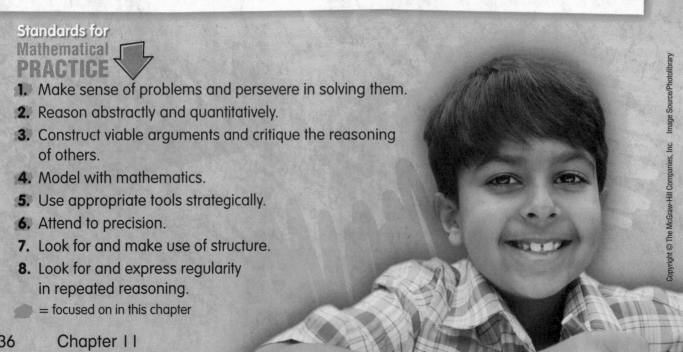

1. Make sense of problems and persevere in solving them.
2. Reason abstractly and quantitatively.
3. Construct viable arguments and critique the reasoning of others.
4. Model with mathematics.
5. Use appropriate tools strategically.
6. Attend to precision.
7. Look for and make use of structure.
8. Look for and express regularity in repeated reasoning.

= focused on in this chapter

Name

Am I Ready?

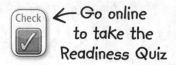 Check ✓ ← Go online to take the Readiness Quiz

Add or subtract.

1.
$$\begin{array}{r} 34 \\ +18 \\ \hline \end{array}$$

2.
$$\begin{array}{r} 26 \\ -17 \\ \hline \end{array}$$

3.
$$\begin{array}{r} 38 \\ +36 \\ \hline \end{array}$$

Measure the length of the object in cubes.

4.

_____ cubes

5.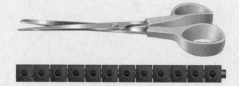

_____ cubes

6. Neil's hand is 5 cubes long. He measures this checkerboard. It is 3 hands long. How many cubes long is the checkerboard?

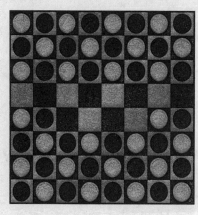

_____ cubes

How Did I Do? ▶ Shade the boxes to show the problems you answered correctly.

| 1 | 2 | 3 | 4 | 5 | 6 |

Name

My Math Words

Vocab abc

Review Vocabulary

compare   longest   shortest

Find three classroom objects. List them. Compare their lengths. Write the longest object. Write the shortest object.

_____   _____   _____

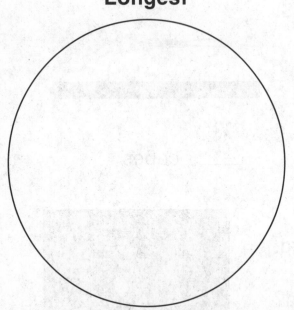

Longest

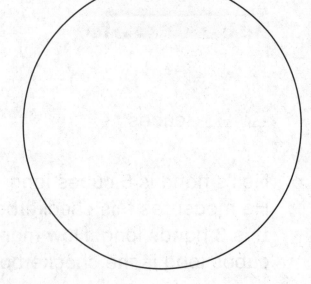

Shortest

How did you compare your objects?

_____

_____

_____

My Vocabulary Cards

 Vocab abc

 Mathematical PRACTICE

Lesson 11–7

centimeter (cm)

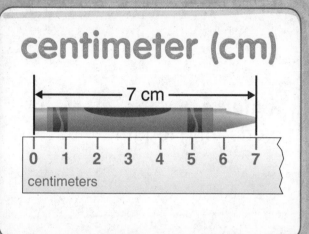

Lesson 11–1

estimate

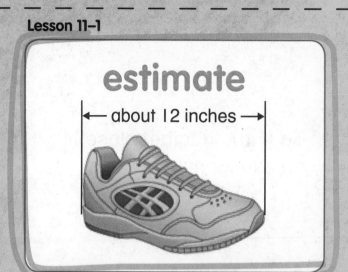

Lesson 11–2

foot (ft)

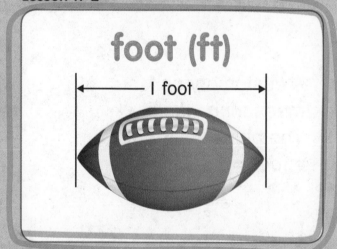

Lesson 11–1

inch (in)

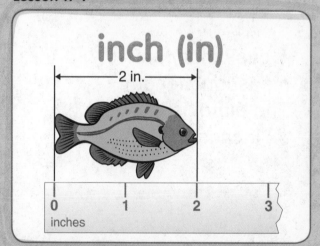

Lesson 11–1

length

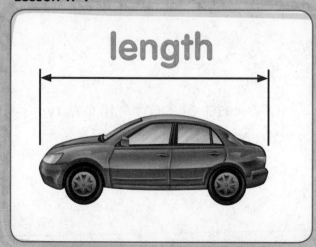

Lesson 11–1

measure

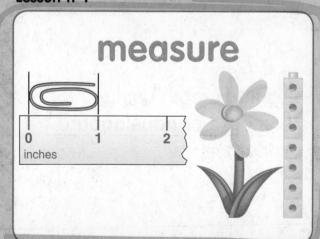

Directions:
Ideas for Use
• Have students group 2 or 3 common words. Ask them to add a word that is unrelated to the group. Have them ask a friend to name the unrelated word.

• Ask students to find pictures to show an example of each word. Have them ask a friend to guess which word the picture shows.

To find a number close to an exact amount.

A metric unit for measuring length.

A customary unit for measuring length.
The plural is inches.
12 inches = 1 foot

A customary unit for measuring length.
The plural is feet.
1 foot = 12 inches

To find length using standard or nonstandard units.

How long or how far away something is.

My Vocabulary Cards

Vocab abc

Mathematical PRACTICE

Lesson 11–7

meter (m)

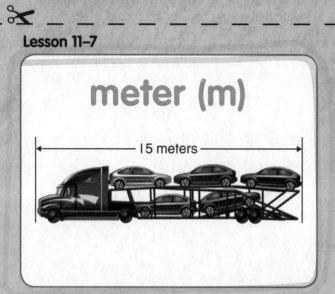

15 meters

Lesson 11–2

yard (yd)

1 yard

Directions:
More Ideas for Use

- Have students use a blank card to write this chapter's essential question. Have them use the back of the card to write or draw examples that help them answer the question.

- Ask students to use the blank cards to write some of the units of measure they learned in this chapter. They should give examples of different objects they would measure with each unit on the back of each card.

A customary unit for measuring length.
1 yard = 3 feet or 36 inches

A metric unit used to measure length.
1 meter = 100 centimeters

My Foldable

FOLDABLES® Follow the steps on the back to make your Foldable.

✂ -

estimate

estimate

yards
inches

centimeters
meters

meters
centimeters

inches
feet

estimate

estimate

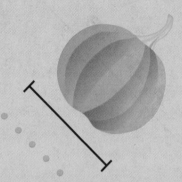

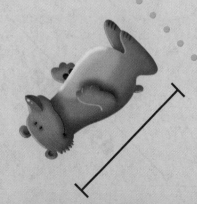

Study Organizer

Name ....

Inches

Lesson 1
ESSENTIAL QUESTION
How can I measure objects?

Explore and Explain

Watch · Tools

Hold still up there!

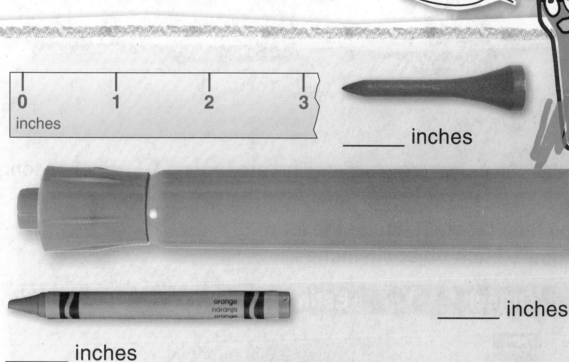

0 1 2 3
inches

_____ inches

_____ inches

_____ inches

_____ inches

Teacher Directions: Place a color tile above the ruler. Line it up with the 0. The color tile is one inch long. Use color tiles to measure the length of each object on the page. Write each length in inches.

See and Show

One **inch** is about the **length** of one color tile. You can use what you know about inches to **estimate** the length of an object. Then use an inch ruler to **measure** the length.

Helpful Hint
Estimate: The length is about 3 inches.

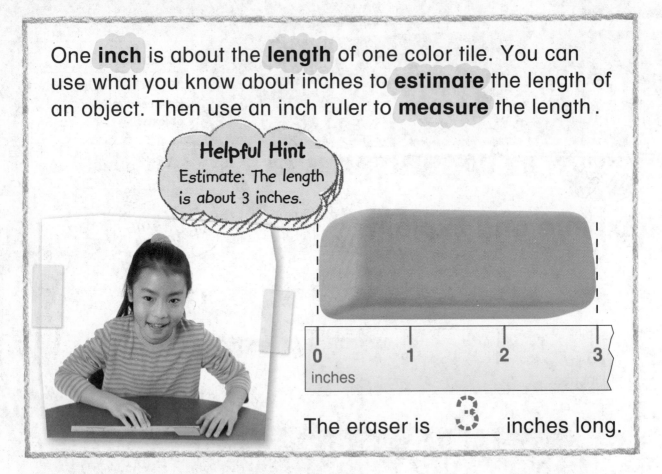

The eraser is _____ inches long.

Find the object. Estimate the length.
Measure each object in inches.

Object	Estimate	Measure
1.	about _____ inches	about _____ inches
2.	about _____ inches	about _____ inches

Talk Math How do you use a ruler to measure inches?

Name

....

On My Own

**Find the object. Estimate the length.
Measure each object in inches.**

Object	Estimate	Measure
3.	about _____ inches	about _____ inches
4.	about _____ inches	about _____ inches
5.	about _____ inches	about _____ inches
6.	about _____ inches	about _____ inches
7.	about _____ inches	about _____ inches

Problem Solving

8. Kaya's surfboard must be at least 15 inches longer than 48 inches. Should Kaya choose a surfboard that is 55 inches long or 65 inches long?

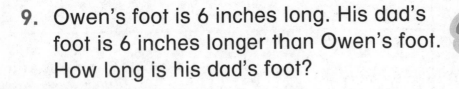

_____ inches long

9. Owen's foot is 6 inches long. His dad's foot is 6 inches longer than Owen's foot. How long is his dad's foot?

_____ inches

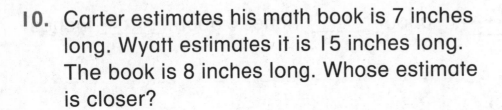

10. Carter estimates his math book is 7 inches long. Wyatt estimates it is 15 inches long. The book is 8 inches long. Whose estimate is closer?

_____

Write Math   How are an estimate and an actual measurement different?

_____

_____

_____

Name

My Homework

Homework Helper

eHelp

Need help? connectED.mcgraw-hill.com

Estimate the length of an object. Then check your estimate by measuring the length with an inch ruler.

Helpful Hint
A small paper clip is about 1 inch long.

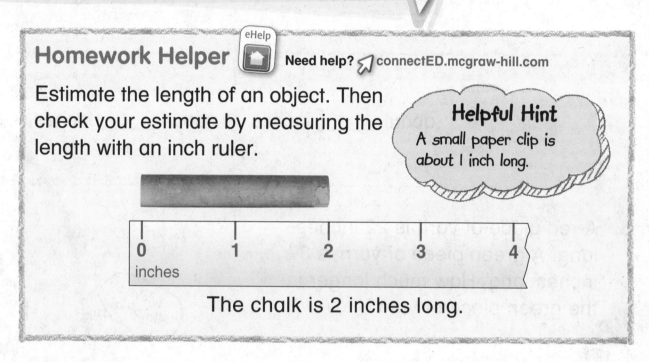

The chalk is 2 inches long.

Practice

**Find the object. Estimate the length.
Measure each object in inches.**

Object	Estimate	Measure
1.	about _____ inches	about _____ inches
2.	about _____ inches	about _____ inches

Find two objects. Draw them. Estimate the length.
Measure each object in inches.

Object	Estimate	Measure
3.	about _____ inches	about _____ inches
4.	about _____ inches	about _____ inches

5. A red piece of yarn is 22 inches
long. A green piece of yarn is 34
inches long. How much longer is
the green piece of yarn?

_____ inches

I love climbing!

Vocabulary Check

Circle the correct answer.

estimate   inch   measure   length

6. An _____ is a unit for measuring length.

 Math at Home Ask your child to measure the length of a fork and a spoon using
an inch ruler.

Measurement and Data
2.MD.1, 2.MD.3, 2.MD.5

Feet and Yards

Lesson 2

ESSENTIAL QUESTION
How can I measure objects?

 Explore and Explain

 Teacher Directions: Discuss each object in the picture. Circle the objects that could be measured with an inch ruler. Draw an X on the objects that are too big to be measured with an inch ruler.

See and Show

You can measure in feet or yards. A **foot** is equal to 12 inches. A **yard** is equal to 36 inches. Length can be measured in any direction.

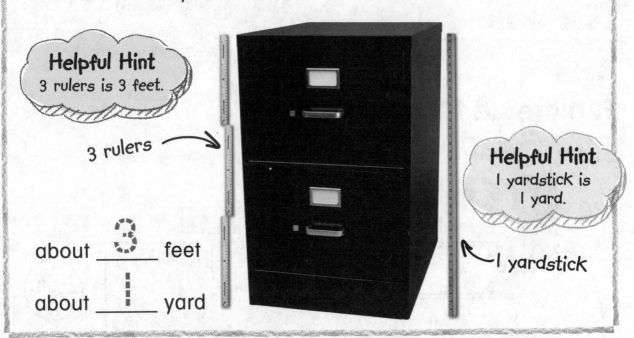

Helpful Hint
3 rulers is 3 feet.

3 rulers

about ___3___ feet

about ___1___ yard

Helpful Hint
1 yardstick is 1 yard.

1 yardstick

**Find the object. Estimate the length.
Measure each object in feet or yards.**

Object	Estimate	Measure
1.	about _____ feet	about _____ feet
2.	about _____ yards	about _____ yards

Talk Math How can you measure a large object with a ruler?

Name _____

On My Own

Find the object. Estimate the length. Measure each object in feet or yards.

Object	Estimate	Measure
3.	about _____ feet	about _____ feet
4.	about _____ feet	about _____ feet
5.	about _____ feet	about _____ feet
6.	about _____ feet	about _____ feet
7.	about _____ yard	about _____ yard

8. The length of Sydney's tennis racket is 2 rulers long. How many feet long is Sydney's tennis racket?

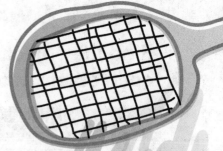

_____ feet

9. Gabriella is 2 inches taller than a yard. How many inches tall is Gabriella?

Sydney has an awesome serve!

_____ inches

10. The pool at the park is 18 feet wide. How many yardsticks would it take to measure the width of the pool?

_____ yardsticks

HOT Problem Josh used 10 yardsticks to measure the length of his basement. How many rulers would he have used? Explain.

_____

_____

_____

Name ....

My Homework

Homework Helper
eHelp

Need help? connectED.mcgraw-hill.com

One foot is equal to 12 inches. One yard is equal to 36 inches. A ruler is 1 foot long. A yardstick is 1 yard long.

Helpful Hint
3 feet = 36 inches

ruler

yardstick

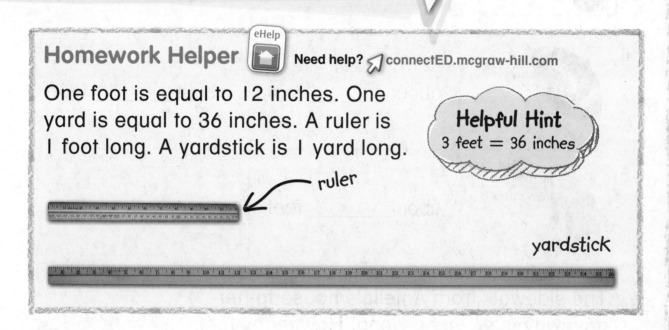

Practice

Find the object. Estimate the length. Measure each object in feet or yards.

Object	Estimate	Measure
1.	about _____ feet	about _____ feet
2.	about _____ yards	about _____ yards

Find the object. Estimate the length. Measure each object in feet or yards.

Object	Estimate	Measure
3.	about _____ yards	about _____ yards
4.	about _____ foot	about _____ foot
5.	about _____ feet	about _____ feet

6. The sidewalk from Ameila's house to her driveway is 96 inches long. How many feet long is the sidewalk?

_____ feet

Sidewalk race!

Hey, wait up!

Vocabulary Check

Circle the correct answer.

7. **foot**

   3 yards   12 inches   36 inches   1 yard

Math at Home Ask your child to measure his or her room using a yardstick or a ruler.

Name ________________________________

Select and Use Customary Tools

Lesson 3

ESSENTIAL QUESTION
How can I measure objects?

Explore and Explain

1.

2.

3.

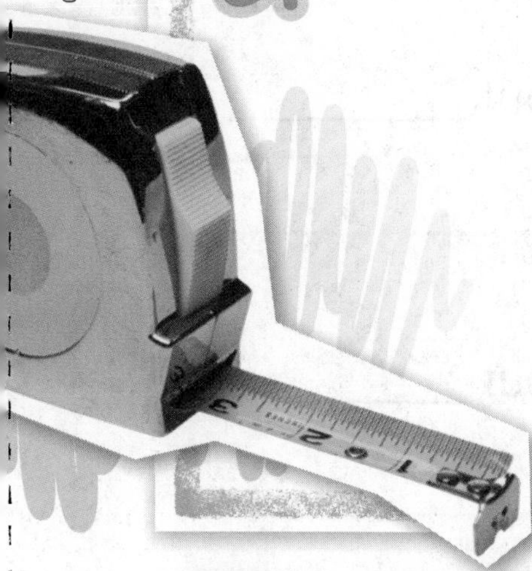

Teacher Directions: Draw an object you would measure with an inch ruler in box 1. Draw an object you would measure with a yardstick in box 2. Draw an object you would measure with a measuring tape in box 3. Explain your drawings.

See and Show

You can select and use tools to measure length. Circle the tool you would use to measure a marker.

Helpful Hint
Measure objects shorter than a foot with an inch ruler, objects longer than a foot with a yardstick, and objects longer than 3 feet with a measuring tape.

← inch ruler

← yardstick

← marker

← measuring tape

Find the object. Choose the tool and measure it. Explain why you chose that tool.

Object	Tool	Measure
1.	_____	about _____
2.	_____	about _____
3.	_____	about _____

Talk Math How do you know which tool to use to measure?

Name

On My Own

**Find the object. Choose the tool and measure it.
Explain why you chose that tool.**

Object	Tool	Measure
4.	_____	about _____
5.	_____	about _____
6.	_____	about _____
7.	_____	about _____
8.	_____	about _____
9.	_____	about _____

Problem Solving

10. Vijay measures the length of his bike with a yardstick. He says it is one yardstick long. How many inches long is it?

_____ inches

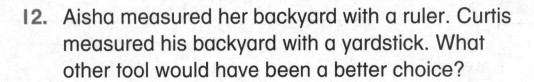

Boy, I'm sweaty after gym class!

11. Ethan measured his foot with a yardstick. What other tool would have been a better choice?

_____

12. Aisha measured her backyard with a ruler. Curtis measured his backyard with a yardstick. What other tool would have been a better choice?

_____

HOT Problem Omar laid 6 yardsticks end to end to measure his driveway. He said it was 6 feet long. Tell why Omar is wrong. Make it right.

_____

_____

_____

Name _____

My Homework

Homework Helper eHelp   **Need help?** connectED.mcgraw-hill.com

You can select and use tools to measure length.

← inch ruler

← yardstick

measuring tape

Helpful Hint

Measure objects shorter than a foot with an inch ruler, objects longer than a foot with a yardstick, and objects longer than 3 feet with a measuring tape.

Practice

Find the object. Choose the tool and measure it.

Object	Tool	Measure
1. 	_____	about _____
2.	_____	about _____

Find the object. Choose the tool and measure it.

Object	Tool	Measure
3.	_____	about _____
4.	_____	about _____
5.	_____	about _____

6. There are 36 inches in 3 feet. How many inches are in 6 feet?

_____ inches

I wish I had 3 feet. I'd go faster!

Test Practice

7. Four feet is equal to _____ inches.

    12   24   36   48
    ◯   ◯   ◯   ◯

Math at Home Have your child measure the width of his or her room using a ruler, a yardstick, and a measuring tape. Talk about which tool is the best choice.

Name _____

Check My Progress

Vocabulary Check

Draw a line to match each word to the correct sentence.

1. **length**   To find a number close to an exact amount.

2. **inch**   Distance from end to end.

3. **estimate**   A customary unit for measuring length.

Concept Check

Find the object. Estimate the length.
Measure each object.

Object	Estimate	Measure
4.	about _____ inches	about _____ inches
5.	about _____ feet	about _____ feet
6.	about _____ yards	about _____ yards

Find the object. Choose the tool and measure it.

Object	Tool	Measure
7.	_____	about _____
8.	_____	about _____
9.	_____	about _____

10. James measured the fence in his backyard with a ruler. It was 24 feet long. Which tool should James have used to measure the fence?

_____

Test Practice

11. Mr. Tom's driveway measures 7 yards long. How many feet long is Mr. Tom's driveway?

7 ○   14 ○   21 ○   28 ○

Name

Compare Customary Lengths

Explore and Explain

I'm taller!

Helpful Hint
Measure length up and down to measure how tall.

Teacher Directions: Find two objects in your classroom. Compare the lengths.
Draw the objects. Explain which object is shorter and which object is longer.

See and Show

You can compare lengths of objects.

baseball bat

__33__ inches

golf club

__31__ inches

The baseball bat is __33__ inches long.

The golf club is __31__ inches long.

The golf club is __2__ inches shorter than the bat.

Think:

$$\begin{array}{r} 33 \\ -31 \\ \hline 2 \end{array}$$

Find the objects. Measure them. Write the lengths.
Write longer or shorter.

1.

_____ inches   _____ inches

The pencil is _____ inches _____ .

2.

_____ feet   _____ feet

The desk is _____ feet _____ .

Talk Math Why do you need to know how
to compare lengths?

Name

On My Own

Find the objects. Measure them. Write the lengths.
Write longer or shorter.

3.

_____ inches   _____inches

The paper clip is _____ inches _____ .

4.

_____ feet   _____ feet

The car is _____ feet _____ .

5.

_____ inches   _____ inches

The shoe is _____ inches _____ .

6.

_____ yards   _____ yards

The door is _____ yards _____ .

Problem Solving

7. The playground is 90 yards long. The field by my house is 80 yards long. How much longer is the playground?

_____ yards

Slapshot!

8. The blue hockey stick is 4 feet long. The red hockey stick is 6 feet long. How much longer is the red hockey stick?

_____ feet

9. Audrey's basketball hoop is 7 feet tall. Isaac's basketball hoop is 10 feet tall. How much taller is Isaac's basketball hoop?

_____ feet

HOT Problem Leah says her classroom is 7 yards wide. Austin says it is 21 feet wide. Both students are correct. Explain.

_____

_____

_____

Name ....

My Homework

Homework Helper eHelp   Need help? connectED.mcgraw-hill.com

You can compare the lengths of objects.

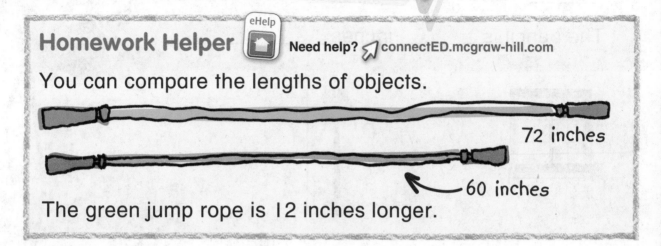

72 inches

60 inches

The green jump rope is 12 inches longer.

Practice

Find the objects. Measure them. Write the lengths. Write longer or shorter.

1.

_____ inches   _____ inches

The crayon is _____ inches _____ .

2.

_____ feet   _____ feet

The rug is _____ feet _____ .

Copyright © The McGraw-Hill Companies, Inc.   (tl)Ingram Publishing/SuperStock, (tr)McGraw-Hill Education, (bl)Judith Collins/Alamy Images, (br)Ingram Publishing/Fotosearch

**Find the objects. Measure them. Write the lengths.
Write longer or shorter.**

3.

_____ inches   _____ inches

The pencil is _____ inches _____.

4.

_____ feet   _____ feet

The window is _____ feet _____.

5. The first grade classroom is 27 feet
wide. The second grade classroom is
24 feet wide. Which classroom is wider?

_____

Test Practice

6. 2 yards is equal to _____ feet.

    3   6   24   32
    ○   ○   ○   ○

 Math at Home Have your child measure two rooms in your house
and compare the lengths.

Name

Relate Inches, Feet, and Yards

Lesson 5

ESSENTIAL QUESTION
How can I measure objects?

Explore and Explain Watch

Knock!
Knock!

Who's there?

_____ inches or _____ feet

_____ inches or _____ feet

_____ inches or _____ feet

 Teacher Directions: Find each object in the classroom. Measure each object in inches. Measure each object in feet. Discuss the measurements.

See and Show

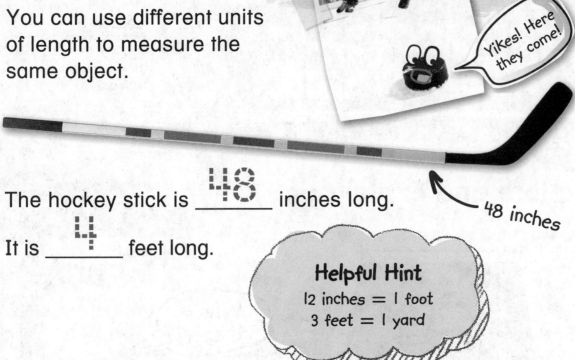

You can use different units of length to measure the same object.

Yikes! Here they come!

The hockey stick is _____48_____ inches long.

It is ____4____ feet long.

48 inches

Helpful Hint
12 inches = 1 foot
3 feet = 1 yard

Find each object. Measure the length of each object twice.

Object	Measure
1.	_____ feet _____ yards
2.	_____ yards _____ feet

Talk Math If there are 12 inches in 1 foot and 3 feet in 1 yard, how many inches are in 1 yard?

Name _____

On My Own

Find each object. Measure the length of each object twice.

Object	Measure
3.	_____ feet _____ inches
4.	_____ inches _____ feet
5.	_____ feet _____ yards
6.	_____ yards _____ feet
7.	_____ feet _____ yards
8.	_____ inches _____ feet

Problem Solving

I rule the pool!

9. Sheradon's scooter is 3 feet long. Sage's scooter is 38 inches long. Whose scooter is longer?

_____

10. The low diving board is 8 feet long. The high diving board is 4 yards long. How many feet longer is the high diving board?

_____ feet

11. The green bike is one yard long. The yellow bike is 38 inches long. Which bike is shorter?

_____ bike

Write Math   Explain how the measurements of an object change depending on which unit you use to measure.

_____

_____

_____

Mathematical
PRACTICE

My Homework

Homework Helper

Need help? connectED.mcgraw-hill.com

You can use different units of length to measure the same object.

The rope is 21 feet long.

It is 7 yards long.

21 feet or 7 yards

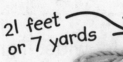

Practice

Find each object. Measure the length of each object twice.

Object	Measurement
1.	_____ feet _____ yards
2.	_____ yards _____ feet
3.	_____ inches _____ feet

Find each object. Measure the length of each object twice.

Object	Measurement
4.	_____ feet _____ inches
5.	_____ feet _____ yards

6. Sam's closet is 4 feet wide. Tom's closet is 50 inches wide. Whose closet is wider?

> **Helpful Hint**
> 12 inches = 1 foot
> 3 feet = 1 yard

_____

7. The fence in Laura's backyard is 15 feet long. How many yards long is the fence?

_____ yards

Test Practice

8. 72 inches is equal to _____ yards.

1	2	3	4
○	○	○	○

 Math at Home Have your child measure an object in your house using inches, feet, and yards. Discuss how each measurement relates to the size of the unit.

Name ....

Problem Solving
STRATEGY: Use Logical Reasoning

Lesson 6
ESSENTIAL QUESTION
How can I measure objects?

Koko wants to plant a garden. She cannot decide if it should be 10 inches, 10 feet, or 100 yards long. About how long should the garden be?

Watch ▶

1 Understand
Underline what you know.
Circle what you need to find.

2 Plan
How will I solve the problem?

3 Solve
Use logical reasoning.

_10_ inches is too short.

_100_ yards is too long.

The garden should be _10_ feet long.

4 Check
Is my answer reasonable? Explain.

Practice the Strategy

Zachary is looking up at the high diving board at the pool. Is the high diving board 10 inches high, 10 feet high, or 10 yards high?

That's my best cannonball EVER!

1 Understand Underline what you know.
Circle what you need to find.

2 Plan How will I solve the problem?

3 Solve I will...

4 Check Is my answer reasonable? Explain.

Apply the Strategy

1. Sam planted a tomato plant that is 1 foot tall. The plant grows a little each week. After 4 weeks, would the plant be 10 inches or 14 inches tall?

_____ inches

2. Jane made a paper chain that is 1 yard long. Brad made a paper chain that is 2 feet long. Who made the longer paper chain?

_____

3. Mr. Moore's class is collecting things to measure. Lisa finds a pinecone. Would the pinecone be 3 inches, 3 feet, or 3 yards long?

_____

Review the Strategies

Choose a strategy
- Use logical reasoning.
- Write a number sentence.
- Make a model.

4. Santos connected 6 markers. Each marker measures 5 inches. How long are all the markers?

_____ inches

5. Jamil had a pencil that was 6 inches long. After a month it was an inch long. How many inches did Jamil use?

I may be short, but I get a lot done!

_____ inches

6. Suni measured from one end of her kitchen table to the middle. It was 42 inches long. Then she measured from the middle to the other end. It was the same length. How long is the table?

_____ inches

Name _____

My Homework

Dave is more than 40 inches tall. He is less than 43 inches tall. His height is an even number. How tall is Dave?

eHelp

Dad says I'm growing like a weed!

1 Understand Underline what you know.
Circle what you need to find.

2 Plan How will I solve the problem?

3 Solve Use logical reasoning.
Dave is either 41 or 42 inches tall.
42 is an even number.
Dave is 42 inches tall.

4 Check Is my answer reasonable?

Problem Solving

Underline what you know. Circle what you need to find. Use logical reasoning to solve.

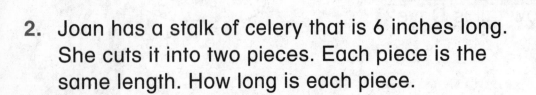

1. A bookshelf is 14 inches wide. There is enough space to fit 2 trophies. One trophy is 2 inches wider than the other. How wide is the wider trophy?

   _____ inches

2. Joan has a stalk of celery that is 6 inches long. She cuts it into two pieces. Each piece is the same length. How long is each piece.

   _____ inches

3. Marlon and Brent are each measuring one of their shoes. Their shoes measure 11 inches in all. Marlon's shoe is 1 inch longer than Brent's shoe. How long is Brent's shoe?

   _____ inches

Test Practice

4. 36 inches is _____ feet.

   1   2   3   4
   ○   ○   ○   ○

Math at Home Ask your child to estimate distances from one room to another. Have our child check by measuring.

Name _____

Check My Progress

Vocabulary Check

Draw lines to match.

1. **measure**

2. **foot**

3. **yard**

To find the length using standard or nonstandard units.

Equal to 36 inches.

Equal to 12 inches.

Concept Check

Find the objects. Measure them. Write the lengths.
Write longer or shorter.

4.

_____ inches   _____ inches

The marker is _____ inches _____.

5.

_____ feet   _____ feet

The cabinet is _____ feet _____.

Find each object. Measure the length of each object twice.

Object	Measure
6. SAVE THE EARTH ↔	_____ feet _____ yards
7.	_____ inches _____ feet
8.	_____ inches _____ feet

9. Bill measured his favorite book with a yardstick.
It was 10 inches long. Which tool should Bill
have used to measure the book?

_____

Test Practice

10. Emily drew a blue line that is 12 inches long.
She drew a purple line that is 4 inches long.
How long are both lines in all?

12 inches   14 inches   16 inches   124 inches
   ○   ○   ○   ○

Name _____

Centimeters and Meters

Lesson 7

ESSENTIAL QUESTION
How can I measure objects?

Explore and Explain ▶Watch

I'm a ping pong pro!

_____ centimeters

_____ centimeters

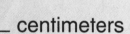

_____ centimeters

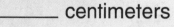

_____ centimeters

Teacher Directions: One unit cube is 1 centimeter long. Use unit cubes to measure each object. Write the length of each object on the line.

See and Show

Mathematical PRACTICE

Use a centimeter ruler to measure in **centimeters**.
Use a meterstick to measure in **meters**. There are
100 centimeters in a meter.

Helpful Hint
Use centimeters to measure
shorter objects. Use meters
to measure longer objects.

centimeters

A paper clip is about

_____5_____ centimeters long.

**Find the object. Estimate the length. Measure
each object in centimeters or meters.**

Object	Estimate	Measure
1.	about _____ centimeters	about _____ centimeters
2.	about _____ meters	about _____ meters
3.	about _____ centimeters	about _____ centimeters

Talk Math Identify objects in the classroom
that are about 1 centimeter long.

Name

On My Own

Find the object. Estimate the length. Measure each object in centimeters or meters.

Object	Estimate	Measure
4.	about _____ centimeters	about _____ centimeters
5.	about _____ centimeters	about _____ centimeters
6.	about _____ centimeters	about _____ centimeters
7.	about _____ meters	about _____ meters
8.	about _____ meters	about _____ meters

Problem Solving

9. Tyler runs the 100 meter dash two times.
   How many meters does he run in all?

   _____ meters

10. Jessa's ski poles are 65 centimeters tall.
    Her sister's ski poles are 80 centimeters tall.
    How much shorter are Jessa's ski poles?

    _____ centimeters

11. Molly's wagon is 100 centimeters long.
    How many meters long is Molly's wagon?

    _____ meter

HOT Problem The length of Isaac's bed
is 2 meters. How many centimeters long is
Isaac's bed? Explain.

_____

_____

_____

Name ....

My Homework

Lesson 7

Centimeters
and Meters

Homework Helper   eHelp   Need help? connectED.mcgraw-hill.com

The cube is 1 centimeter long.
There are 100 centimeters in
1 meter.

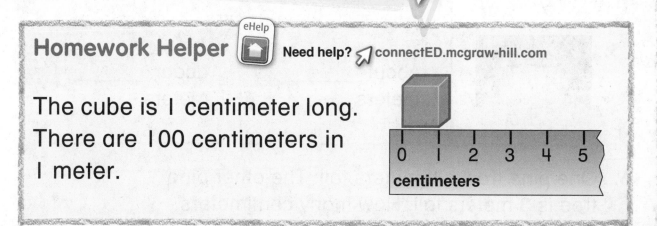

Practice

**Find the object. Estimate the length. Measure
each object in centimeters or meters.**

Object	Estimate	Measure
1.	about _____ centimeters	about _____ centimeters
2.	about _____ centimeters	about _____ centimeters
3.	about _____ meters	about _____ meters

Find the object. Estimate the width. Measure each object in centimeters or meters.

Object	Estimate	Measure
4.	about _____ centimeters	about _____ centimeters
5.	about _____ meters	about _____ meters

6. One pine tree is 4 meters tall. The other pine tree is 3 meters tall. How many centimeters taller is the first pine tree?

_____ centimeters

Vocabulary Check

7. Circle the answer that is the same as 1 **meter**.

   1 centimeter   10 centimeters

   100 centimeters   1,000 centimeters

Math at Home Have your child identify objects he or she would measure using centimeters.

Name
....

Select and Use Metric Tools

Lesson 8

ESSENTIAL QUESTION
How can I measure objects?

I will now perform a half-twist!

Explore and Explain Watch ▶

1.

2.

Teacher Directions: Draw a picture of an object you would measure with a centimeter ruler in box 1. Draw a picture of an object you would measure with a meterstick in box 2.

See and Show

A centimeter ruler measures smaller objects.
A meterstick measures larger objects.
Circle the tool you would use to measure
a lunchbox.

Lunch time

centimeter ruler

meterstick

Find the object. Choose the tool and measure it.
Explain why you chose that tool.

Object	Tool	Measure
1.	_____	about _____
2.	_____	about _____
3.	_____	about _____

Talk Math Can you measure a paper clip
with a meterstick? Explain.

Name _____

On My Own

Find the object. Choose the tool and measure it.
Explain why you chose that tool.

Object	Tool	Measure
4.	_____	about _____
5.	_____	about _____
6.	_____	about _____
7.	_____	about _____
8.	_____	about _____

9. A wall in Jim's garage is 5 meters long. He is painting 2 meters of the wall blue. He wants to paint the rest of the wall red. How many meters will he paint red?

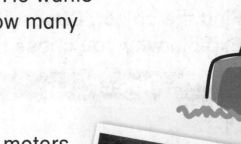

_____ meters

10. A sandbox is 300 centimeters long. How many meters long is it?

_____ meters

11. Allison swam 50 meters in the morning. She swam 40 meters in the evening. How many meters did Allison swim in all?

_____ meters

Write Math Why do you need to know customary and metric length?

_____

_____

_____

My Homework

Homework Helper
eHelp

Need help? connectED.mcgraw-hill.com

A centimeter ruler is used to measure smaller objects.

centimeter ruler

A meterstick is used to measure longer objects.

meterstick

Practice

**Find the object. Choose the tool and measure it.
Explain why you chose that tool.**

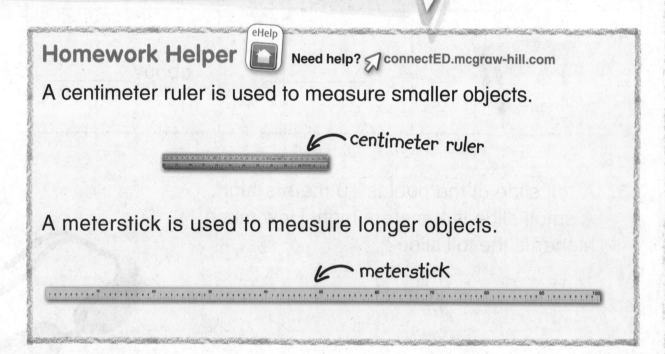

Object	Tool	Measure
1.	_____	about _____
2.	_____	about _____

Find the object. Choose the tool and measure it.
Explain why you chose that tool.

Object	Tool	Measure
3.	_____	about _____
4.	_____	about _____

5. A tall slide at the pool is 10 meters high.
A small slide is 4 meters high. How much
higher is the tall slide?

_____ meters

6. The hundred meter dash is marked with
a blue line. The 400 meter dash is marked
with a red line. How much farther is the red line?

_____ meters

Test Practice

7. Eight meters is equal to _____ centimeters.

   400   600   800   900
    ○   ○   ○   ○

Math at Home Have your child identify objects he or she would measure
using meters and centimeters.

Measurement and Data

2.MD.4

Compare Metric Lengths

Lesson 9
ESSENTIAL QUESTION
How can I measure objects?

Explore and Explain

Measuring centimeters is a KICK!

_____ centimeters   _____ centimeters

Teacher Directions: Find two objects in your classroom. Draw a picture of each object. Measure each object in centimeters. Write how many centimeters long each object is. Circle the object that is longer.

See and Show

You can compare lengths of objects.

69 centimeters

76 centimeters

The blue tennis racket is ___69___ centimeters long.

The green tennis racket is ___76___ centimeters long.

The blue racket is ___7___ centimeters shorter than the green racket.

Find the objects. Measure them. Write the lengths.
Write longer or shorter.

1.

_____ centimeters   _____ centimeters

The lunchbox is _____ centimeters _____.

2.

_____ meters   _____ meters

The car is _____ meters _____.

Name ....

On My Own

Find the objects. Measure them. Write the lengths.
Write longer or shorter.

3.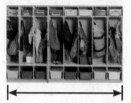

_____ meters   _____ meters

The slide is _____ meters _____ .

4.

_____ centimeters   _____ centimeters

The knife is _____ centimeters _____ .

5.

_____ meters   _____ meters

The swing set is _____ meters _____ .

6.

_____ centimeters   _____ centimeters

The hair brush is _____ centimeters _____ .

Problem Solving

7. The red skateboard is 75 centimeters long. The purple skateboard is 83 centimeters long. How much longer is the purple skateboard?

_____ centimeters

8. Marcus ran 50 meters. Will ran 67 meters. How much farther did Will run?

_____ meters

9. Alec rode his bike 82 meters. Kyla rode her bike 14 meters farther than Alec. How far did Kyla ride her bike?

_____ meters

Write Math   How can you compare a measurement in centimeters to a measurement in meters?

_____

_____

My Homework

Lesson 9

Compare Metric Lengths

Homework Helper Need help? connectED.mcgraw-hill.com

You can compare lengths of objects.

10 centimeters   7 centimeters

The brown guinea pig is 3 centimeters longer.

Practice

**Find the objects. Measure them. Write the lengths.
Write longer or shorter.**

1.

_____ centimeters

_____ centimeters

The spoon is _____ centimeters _____.

2.

_____ meters

_____ meters

The house is _____ meters _____.

**Find the objects. Measure them. Write the lengths.
Write longer or shorter.**

3.    

_____ centimeters   _____ centimeters

The crayon is _____ centimeters _____ .

4. 

_____ meters   _____ meters

The couch is _____ meters _____ .

> How far is the house from here?

5. The path from my house to the barn
is 50 meters. The path from the barn
to the pond is 30 meters. How long is
the path in all?

_____ meters

Test Practice

6. 56 meters + 35 meters = _____ meters.

     86   87   91   95
     ○   ○   ○   ○

 Math at Home Have your child measure two beds in your house and compare
the lengths in meters.

Name ....

Relate Centimeters and Meters

Lesson 10

ESSENTIAL QUESTION
How can I measure objects?

Explore and Explain

_____ centimeters or _____ meters

Teacher Directions: Choose a large object from the picture such as a table or window. Find it in your school. Measure the object in centimeters. Measure the object in meters. Discuss the measurements.

Online Content at 🖱 **connectED.mcgraw-hill.com**

Chapter 11 • Lesson 10

See and Show

You can use different units of length to measure the same object.

The basketball hoop is about

_____**3**_____ meters tall.

It is about ____**300**____ centimeters tall.

Want to see my "behind-the-back, nothing but net" shot?

Helpful Hint
1 meter = 100 centimeters

Find the object. Measure the length of each object twice.

Object	Measure
1.	_____ centimeters _____ meters
2.	_____ centimeters _____ meters
3.	_____ centimeters _____ meters

Talk Math Which unit of measure gives you a more exact measurement?

Name ....

On My Own

Find the object. Measure the length of each object twice.

Object	Measure
4.	_____ centimeters _____ meters
5.	_____ centimeters _____ meters
6.	_____ centimeters _____ meters
7.	_____ centimeters _____ meters
8.	_____ centimeters _____ meters

Problem Solving

9. Kate's garden is 9 meters long. Stella's garden is 954 centimeters long. Whose garden is longer?

_____

10. Brianna's bedroom is 6 meters wide. Oliver's bedroom is 596 centimeters wide. Whose bedroom is wider?

I'm stuffed!

_____

11. Gianna walks 10 meters to her mailbox. Parker walks 757 centimeters to his mailbox. Who walks farther?

_____

HOT Problem Abby's ribbon was 5 meters long. She cut off 35 centimeters. How long is the piece of ribbon now? Explain.

_____

_____

_____

_____

Name

My Homework

Homework Helper **Need help?** connectED.mcgraw-hill.com

You can use different units of length to measure the same object.

The yarn is 2 meters long.
It is 200 centimeters long.

Practice

Find the object. Measure the length of each object twice.

Object	Measure
I.	_____ centimeters _____ meters
2.	_____ centimeters _____ meters
3.	_____ centimeters _____ meters

Find the object. Measure the length of each object twice.

Object	Measure
4.	_____ centimeters _____ meters
5.	_____ centimeters _____ meters

6. At the end of the race, Carly was 3 meters ahead of Reagan. How many centimeters was she ahead of Reagan?

3 is my lucky number!

_____ centimeters

Test Practice

7. 35 centimeters + 65 centimeters = _____

I centimeter	I meter	90 centimeters	2 meters
○	○	○	○

Math at Home Have your child measure an object in your house using centimeters and meters. Discuss how each measurement relates to the size of the unit.

Measure on a Number Line

Lesson 11

ESSENTIAL QUESTION
How can I measure objects?

Explore and Explain

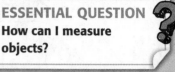

I wonder where I'd be on a number line?

centimeters

0 1 2 3 4 5 6 7 8 9 10 11 12 13 14 15

_____ − _____ = _____

Teacher Directions: Place a crayon and a pair of scissors on the number line. Line up the ends at 0. How much longer are the scissors than the crayon? Repeat with other pairs of objects. Draw a pair of objects on the number line and write the subtraction sentence that shows how much longer one object is than another.

See and Show

centimeters

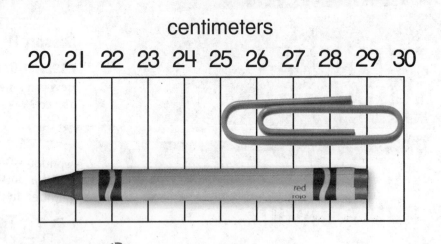

The crayon is ___9___ centimeters long.

The paper clip is ___5___ centimeters long.

Use the number line to answer the questions.

inches

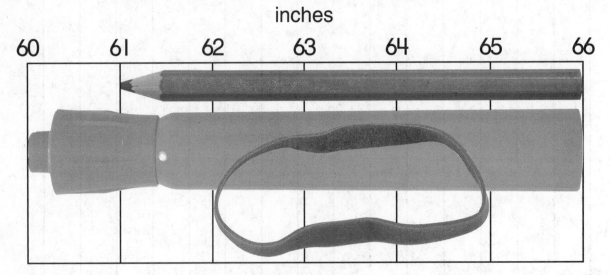

1. How long is the pencil? _____ inches

2. How much longer is the marker than
   the rubber band? _____ inches

Talk Math How does the number line
help you compare measurements?

Name _____

On My Own

Use the number line to answer the questions.

centimeters

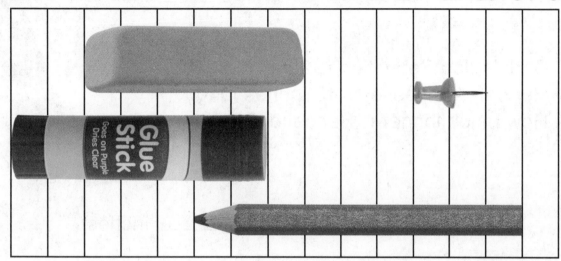

70 71 72 73 74 75 76 77 78 79 80 81 82 83 84 85

3. How long is the glue stick?   _____ centimeters

4. How long is the push pin?   _____ centimeters

5. How long is the eraser?   _____ centimeters

6. How much longer is the pencil
    than the glue stick?   _____ centimeters

7. How long are the eraser and
    push pin in all?   _____ centimeters

8. How much shorter is the push
    pin than the pencil?   _____ centimeters

Problem Solving

Mathematical PRACTICE

9. Makenna is 38 inches tall. Her brother is 26 inches tall. How much taller is Makenna?

See, I told you I was longer!

_____ inches

10. Carly's dog Sebastian is 22 inches long. Her cat, Annabelle, is 17 inches long. How much longer is Sebastian?

_____ inches

Meow!

11. Levi found a trail in the woods that is 54 meters long. Then he found a second trail that is 83 meters long. How much longer is the second trail?

_____ meters

How can a number line help you measure objects? Explain.

_____

_____

_____

Copyright © The McGraw-Hill Companies, Inc.   GK Hart/Vikki Hart/Photodisc/Getty Images

Name _____

My Homework

Homework Helper

eHelp

Need help? connectED.mcgraw-hill.com

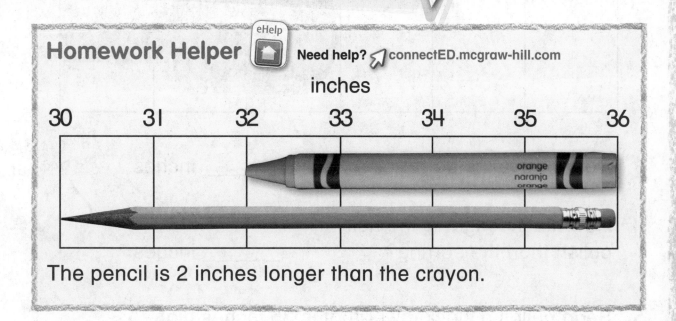

inches

30   31   32   33   34   35   36

The pencil is 2 inches longer than the crayon.

Practice

Use the number line to answer the questions.

centimeters

60 61 62 63 64 65 66 67 68 69 70 71 72 73 74 75

1. How long is the truck?   _____ centimeters

2. How long is the car?   _____ centimeters

Use the number line to answer the questions.

inches

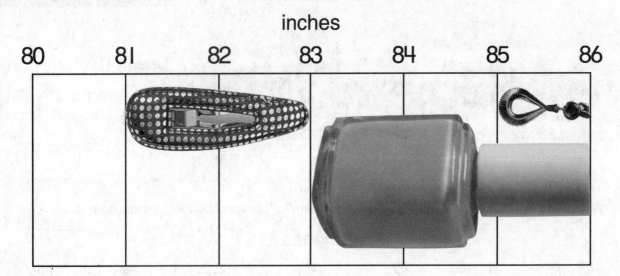

80   81   82   83   84   85   86

3. How long is the barrette?   _____ inches

4. How much longer is the nail
   polish than the earring?   _____ inches

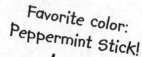

Favorite color:
Peppermint Stick!

5. Sean builds a block tower that is 35 inches high.
   Victoria builds a block tower that is 27 inches
   high. How much higher is Sean's tower?

_____ inches

Test Practice

6. 36 inches + 26 inches = _____

   62 inches   62 centimeters   62 feet   62 meters
     ○   ○   ○   ○

Math at Home Help your child create a large number line measured
in inches. Have him or her measure objects from your home using the
number line.

Name ....

Measurement Data

Explore and Explain

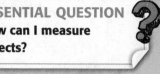
I can measure anything!

I am "Green Ruler Man"!

Pencil Lengths

0 1 2 3 4 5 6 7 8 9 10 11 12
inches

Teacher Directions: Measure five of your classmates' pencils with a ruler.
Make a line plot using the data from your measurements.

Copyright © The McGraw-Hill Companies, Inc.   Medioimages/Photodisc/Getty Images

See and Show

Helpful Hint
Remember a line plot tells how many times a number occurs in data.

Shoe Lengths

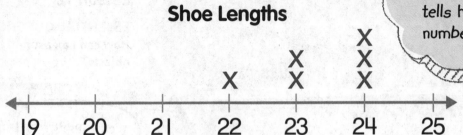

19   20   21   22   23   24   25
centimeters

How many people have shoes that are 23 centimeters long? ___2___

Measure the length of the right hand of 10 people in inches. Use the data to make a line plot.

Hand Lengths

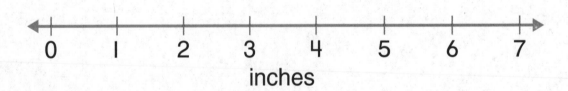

0   1   2   3   4   5   6   7
inches

1. How long are most of the hands?   _____ inches

2. How long is the longest hand?   _____ inches

Talk Math How does a line plot help to show measurement data?

Name _____

On My Own

**Measure 15 books from your classroom library
in centimeters. Use the data to make a line plot.**

Book Lengths

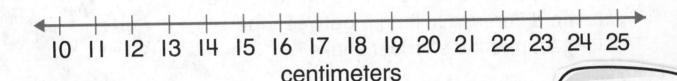

10 11 12 13 14 15 16 17 18 19 20 21 22 23 24 25
centimeters

> This one's a whopper!

3. How long are most books?   _____ centimeters

4. How long is the longest book? _____ centimeters

**Measure 15 used crayons in inches.
Use the data to make a line plot.**

Crayon Lengths

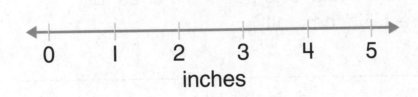

0   1   2   3   4   5
inches

5. How short is the shortest crayon? _____ inches

6. How long is the longest crayon? _____ inches

Problem Solving

Mathematical
PRACTICE

7. A line plot shows that 12 people are 48 inches tall, 8 people are 50 inches tall, and 9 people are 47 inches tall. How tall are most of the people?

Did they count my tongue?

_____ inches

8. Four snakes are 20 inches, 2 snakes are 32 inches, and 4 snakes are 15 inches. What is the difference in length between the longest and the shortest snake?

_____ inches

9. Four people went down the slide 8 times. Eight people went down the slide 6 times. Two people went down the slide 12 times. How many times did most people go down the slide?

_____ times

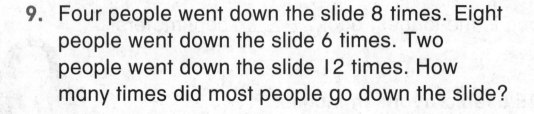

HOT Problem Jordan made a line plot of the heights of his 5 pets. The same number of pets are 14 inches and 18 inches tall. One pet is 6 inches tall. How many pets are 14 inches tall?

_____

_____

_____

Name ....

My Homework

Homework Helper Need help? connectED.mcgraw-hill.com

Three people have pencils that are 8 centimeters long.

Two people have pencils that are 13 centimeters long.

Pencil Lengths

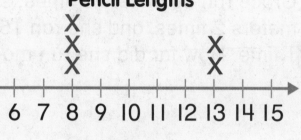

centimeters

Practice

Measure the thumb length of each of your family members. Use the data to make a line plot.

Helpful Hint
Line plots show how often a number occurs in data.

Thumb Lengths

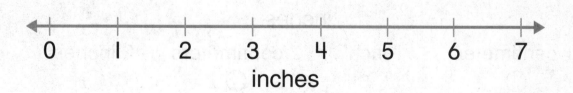

inches

1. How long is the longest thumb? _____ inches

2. How short is the shortest thumb? _____ inches

Chapter 11 • Lesson 12   719

3. Make a line plot to
   show how tall each
   flower is. Four flowers
   are 10 centimeters,
   3 flowers are
   13 centimeters,
   and 6 flowers are
   17 centimeters.

10 11 12 13 14 15 16 17 18 19 20

4. Grace ran 50 meters 3 times, she ran 100
   meters 2 times, and she ran 150 meters
   1 time. How far did she run most of the time?

_____ meters

Test Practice

5. Use the line plot below to answer the question.
   How long are most toes?

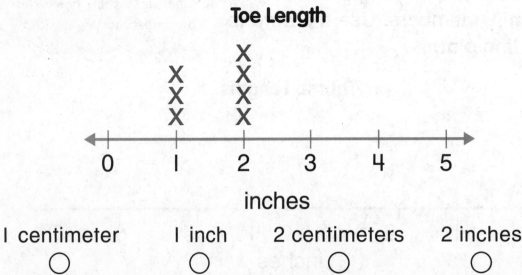

Toe Length

inches

○ 1 centimeter   ○ 1 inch   ○ 2 centimeters   ○ 2 inches

Math at Home Have your child create a line plot to show the heights
of everyone in your family.

Name ....

My Review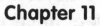

Vocabulary Check

Write the correct word in each blank.

| length | inch | estimate | measure |
| foot | yard | centimeter | meter |

1. You can _____ to find a number close
   to an exact amount.

2. _____ is how long or how far away
   something is.

3. A paper clip is about one _____
   long.

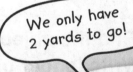

We only have
2 yards to go!

4. A _____ is equal to 3 feet.

5. To _____ is to find the length, height,
   or weight using standard or nonstandard units.

6. A base-ten cube is about one _____ long.

7. A _____ is equal to 12 inches.

8. A _____ is equal to 100 centimeters.

Concept Check ✓

**Find the object. Estimate the length. Measure
the object in centimeters.**

Object	Estimate	Measure
9.	about _____ centimeters	about _____ centimeters

Find the object. Choose the tool and measure it.

Object	Tool	Measure
10.	_____	about _____

**Find the objects. Measure them. Write the lengths.
Write longer or shorter.**

11.

_____ centimeters   _____ centimeters

The glue stick is _____ centimeters _____.

Find the object. Measure the length of it twice.

Object	Measure
12.	_____ centimeters _____ meters

Name _____

 Problem Solving

13. Use the data to make a line plot.
6 pieces of yarn were 4 inches long,
3 pieces of yarn were 2 inches long,
2 pieces of yarn were 6 inches long,
and 1 piece of yarn was 5 inches long.

Yarn Lengths

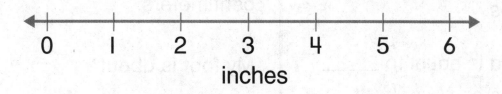

0   1   2   3   4   5   6

inches

---

14. Mason's bike is 4 feet long. His scooter is
2 feet long. How many inches longer is his
bike than his scooter?

_____ inches

Test Practice

15. Lucy's lacrosse stick is 1 meter long. How many
centimeters long is Lucy's lacrosse stick?

50   100   200   400
○   ○   ○   ○

Reflect

Complete each sentence.

ESSENTIAL QUESTION How can I measure objects?

Customary	Metric
A foot is equal to _____ inches.	A meter is equal to _____ centimeters.
A yard is equal to _____ inches or _____ feet.	My foot is about _____ centimeters long.
My foot is about _____ inches long.	I can measure metric lengths with a _____ or a _____.
I can measure customary lengths with a _____, _____, or _____.	

We're good sports too!

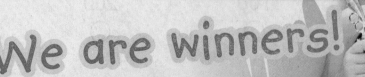

We are winners!

Chapter 12

Geometric Shapes and Equal Shares

ESSENTIAL QUESTION
How do I use shapes and equal parts?

Let's Go to the Park!

Watch a video!

Watch

My Standards

Geometry

2.G.1 Recognize and draw shapes having specified attributes, such as a given number of angles or a given number of equal faces. Identify triangles, quadrilaterals, pentagons, hexagons, and cubes.

2.G.2 Partition a rectangle into rows and columns of same-size squares and count to find the total number of them.

2.G.3 Partition circles and rectangles into two, three, or four equal shares, describe the shares using the words *halves, thirds, half of, a third of,* etc., and describe the whole as two halves, three thirds, four fourths. Recognize that equal shares of identical wholes need not have the same shape.

Standards for Mathematical PRACTICE

1. Make sense of problems and persevere in solving them.
2. Reason abstractly and quantitatively.
3. Construct viable arguments and critique the reasoning of others.
4. Model with mathematics.
5. Use appropriate tools strategically.
6. Attend to precision.
7. Look for and make use of structure.
8. Look for and express regularity in repeated reasoning.

= focused on in this chapter

Go online to take the Readiness Quiz

Draw an X over the object that is a different shape.

1.    2.

Draw a line to match the objects that are the same shape.

3.

4.

5.

6. What shape is the game board?
   Circle the word.

   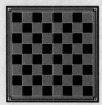

   triangle   square   circle

How Did I Do?

Shade the boxes to show the problems you answered correctly.

1	2	3	4	5	6

My Math Words

Vocab
abc

Review Vocabulary

circle   rectangle   square

Use the review vocabulary to complete the chart.
The first row is done for you.

Word	Example	Not an Example
triangle	△	○
_____	□	
_____	○	
_____	▭	

My Vocabulary Cards

Lesson 12-2

angle

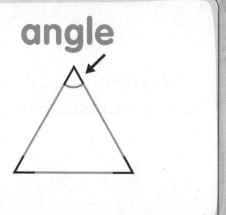

Lesson 12-4

cone

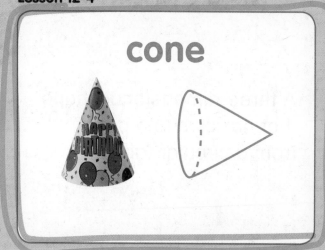

Lesson 12-4

cube

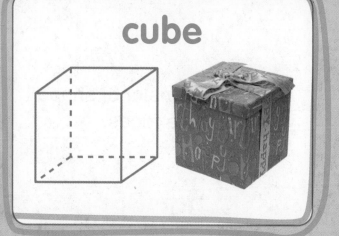

Lesson 12-4

cylinder

Lesson 12-5

edge

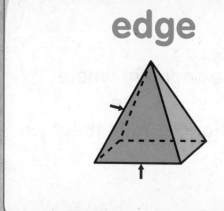

Lesson 12-5

face

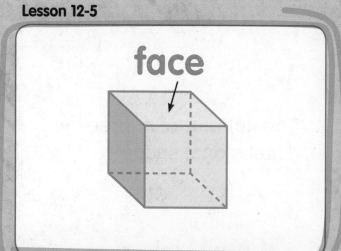

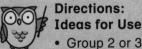

- Group 2 or 3 common words. Add a word that is unrelated to the group. Ask another student to name the unrelated word.

- Have students draw an example for each card. Have them make drawings that are different from what is shown on each card.

A three-dimensional shape that narrows to a point from a circular face.

Where two sides on a two-dimensional shape meet.

A three-dimensional shape that is shaped like a can.

A three-dimensional shape with 6 square faces.

The flat part of a three-dimensional shape.

The line segment where two faces of a three-dimensional shape meet.

My Vocabulary Cards

 Vocab

Lesson 12-7

fourths

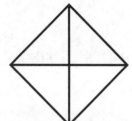

4 fourths

Lesson 12-7

halves

2 halves

Lesson 12-1

hexagon

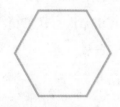 

Lesson 12-1

parallelogram

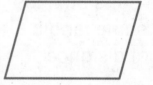 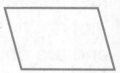

Lesson 12-7

partition

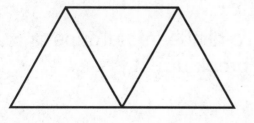

Lesson 12-1

pentagon

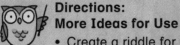

Directions:
More Ideas for Use

- Create a riddle for each word. Have a friend guess the word for each riddle.

- Have students write a tally mark on each card every time they read the word in this chapter or use it in their writing. Challenge them to try to use at least 10 tally marks for each word card.

Two equal parts.

Four equal parts.

A two-dimensional shape that has two pairs of sides that are the same length and are equal distance apart.

A two-dimensional shape that has six sides.

A polygon with five sides.

To divide into groups or "break up."

My Vocabulary Cards

Lesson 12-4

pyramid

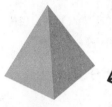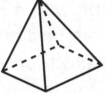

Lesson 12-1

quadrilateral

Lesson 12-2

rectangular prism

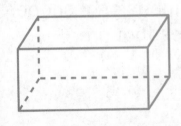

Lesson 12-4

side

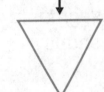

Lesson 12-6

sphere

Lesson 12-1

thirds

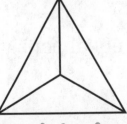

3 thirds

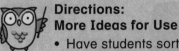

Directions:
More Ideas for Use
- Have students sort the words by the number of syllables in each word.

- Ask students to find pictures to show each word. Have them work with a friend to guess which word the picture shows.

A shape that has 4 sides and 4 angles.

A three-dimensional shape with a polygon as a base and other faces that are triangles.

One of the line segments that make up a shape.

A three-dimensional shape with 6 faces that are rectangles.

Three equal parts.

A three-dimensional shape that has the shape of a round ball.

My Vocabulary Cards

Lesson 12-4

three-dimensional shapes

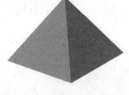

Lesson 12-7

trapezoid

Lesson 12-1

two-dimensional shapes

Lesson 12-1

vertex

vertex

Directions:
More Ideas for Use
- Have students arrange the cards in alphabetical order.

- Group 2 or 3 common words. Add a word that is unrelated to the group. Ask another student to name the unrelated word.

A two-dimensional shape with four sides and only two opposite sides that are parallel.

A shape that has: length, width, and height.

The point where three or more faces of a three-dimensional shape meet.

A shape having the two dimensions length and width.

My Foldable

 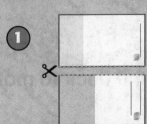 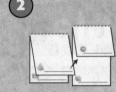

Name ....

Two-Dimensional Shapes

Lesson 1

ESSENTIAL QUESTION
How do I use shapes and equal parts?

Explore and Explain

Don't be such a square.

circle

hexagon

square

rectangle

triangle

Teacher Directions: Use small attribute blocks. Trace and identify each shape.
Draw a line from each shape to its name.

See and Show

A **two-dimensional shape** is a shape with only length and width.

circle   **triangle**   **square**   **rectangle**

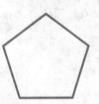 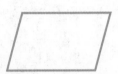 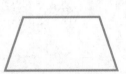

pentagon   **hexagon**   **parallelogram**   **trapezoid**

Circle the shapes that match the name.

1. parallelogram

2. triangle

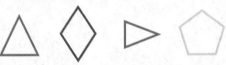

Write the name of the shape.
Circle the shape that matches.

3.

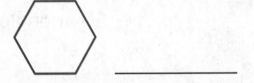 _____

Talk Math   What is the difference between a pentagon and a hexagon? How are they alike?

Name

On My Own

Circle the shapes that match the name.

4. trapezoid

 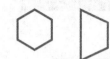

5. hexagon

6. triangle

7. pentagon

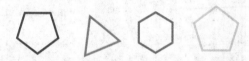

Write the name of the shape.
Circle the shape that matches.

8.

 _____

9.

 _____

Circle the shape that does not belong in each group.

10.

11.

Problem Solving

12. Identify the shape of each sign.

_____   _____

How many of each shape do you see?

13.

City Zoo

Polar Bears	Bears
Big Cats	Food Court

Penguins

Front Gate

Where is my house?

Monkeys

Apes

Hippos

Elephants

Eagles

Otters

Giraffes

triangles _____   hexagons _____   rectangles _____

squares _____   pentagons _____   circles _____

Write Math Give examples of objects in your school that look like triangles and squares.

_____

_____

_____

Name _____

My Homework

Lesson 1
Two-Dimensional Shapes

Homework Helper Need help? connectED.mcgraw-hill.com

A two-dimensional shape is a shape with only length and width.

circle

triangle

square

rectangle

pentagon

hexagon

parallelogram

trapezoid

Practice

Circle the shapes that match the name.

1. rectangle

2. triangle

3. trapezoid

4. hexagon

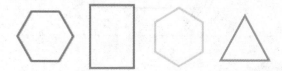

Write the name of the shape. Circle the shape that matches.

5.

_____

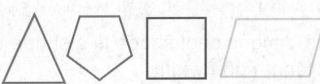

6.

_____

7. Jack cut out a shape to glue onto a picture. The shape looked like an ice cream cone. What shape did he cut out?

_____

mmm, strawberry!

Vocabulary Check

8. Circle the **hexagons**.

Image Source/Getty Images

Copyright © The McGraw-Hill Companies, Inc.

Math at Home Point to two-dimensional shapes around your house (triangles, squares, rectangles, hexagons, and pentagons) and have your child identify each shape.

Name ....

Sides and Angles

Lesson 2

ESSENTIAL QUESTION
How do I use shapes and equal parts?

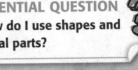

Explore and Explain 　Watch ▶ 　Tools

Did you see what I saw?

_____ _____
sides angles

_____ _____
sides angles

_____ _____
sides angles

 Teacher Directions: Have students sort triangle, square, parallelogram, trapezoid, and hexagon pattern blocks by their number of sides and angles. Trace them. Write how many sides and angles.

Online Content at 　connectED.mcgraw-hill.com

See and Show

Mathematical
PRACTICE

You can describe two-dimensional shapes by the number of **sides** and **angles**.

triangle

side
angle

_____ sides

_____ angles

quadrilateral

_____ sides

_____ angles

pentagon

_____ sides

_____ angles

hexagon

_____ sides

_____ angles

circle

_____ sides

_____ angles

Trace each shape. Write how many sides and angles.

1. _____ sides

_____ angles

2. _____ sides

_____ angles

3. Circle the objects that have 0 sides and 0 angles.

Talk Math How are a square and a hexagon alike? How are they different?

Name

On My Own

Trace each shape. Write how many sides and angles.

4.

_____ sides

_____ angles

5.

_____ sides

_____ angles

6.

_____ sides

_____ angles

7.

_____ sides

_____ angles

Circle the objects that match the description.

8. 3 sides and 3 angles

9. 4 sides and 4 angles

Problem Solving

Draw a picture to solve.

10. Kira draws a shape with 6 sides and 6 angles. What shape does she draw?

_____

11. Alex draws a shape with 3 sides and 3 angles. What shape does Alex draw?

_____

12. Josh drew 3 squares. Katie drew 2 triangles and 1 square. Who drew more angles?

_____

Write Math   Write the name of each shape. Describe two things about each shape.

⬠ _____

1. _____

2. _____

□ _____

1. _____

2. _____

Name

My Homework

Lesson 2

Sides and Angles

Homework Helper 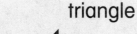 Need help? connectED.mcgraw-hill.com

A two-dimensional shape can be described
by its sides and angles.

triangle

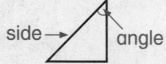 side → angle

3 sides
3 angles

quadrilateral

4 sides
4 angles

pentagon

5 sides
5 angles

hexagon

6 sides
6 angles

circle

0 sides
0 angles

Practice

Trace each shape. Write how many sides and angles.

1.

_____ sides

_____ angles

2.

_____ sides

_____ angles

3.

_____ sides

_____ angles

4.

_____ sides

_____ angles

5. Circle the object that has 8 sides and 8 angles.

6. Jason drew a shape that has 6 sides. What shape did he draw?

_____

7. Carla drew a triangle and a square. Janice drew a shape with 6 sides and 6 angles. Who drew more sides and angles?

_____

Circle time!

Vocabulary Check

Connect the name of each shape to its number of sides or angles.

8. **hexagon**   4 sides and 4 angles

9. **quadrilateral**   5 sides and 5 angles

10. **triangle**   6 sides and 6 angles

11. **pentagon**   3 sides and 3 angles

Math at Home While driving or walking, look at road signs together. Ask your child to name and describe the shapes of the signs he or she sees.

Problem Solving
STRATEGY: Draw a Diagram

Lesson 3

ESSENTIAL QUESTION
How do I use shapes
and equal parts?

Lyla drew a shape. The shape
has 6 sides. It also has 6 angles.
What shape did Lyla draw?

Watch Tools

This way to the
park soccer field.

I have 6 sides!

I have 5!

1 Understand Underline what you know.
Circle what you need to find.

2 Plan How will I solve the problem?

3 Solve Draw a diagram.

Lyla drew a _____.

4 Check Is my answer reasonable? Explain.

Practice the Strategy

Marcy drew a shape.
It has 5 sides. It has 5 angles.
What shape did she draw?

1 Understand Underline what you know.
Circle what you need to find.

2 Plan How will I solve the problem?

3 Solve I will...

Marcy drew a _____.

4 Check Is my answer reasonable? Explain

Drawing at the park is fun!

Name

Apply the Strategy

1. If a shape has 3 sides and 3 angles, what shape is it? Draw the shape.

   _____

2. Abby draws a triangle. Samuel draws a shape that has 1 more side than a triangle. What shape did Samuel draw? Draw the shape.

   _____

3. Jason drew a shape that has more sides than a triangle or rectangle but less angles than a hexagon. What shape did he draw? Draw the shape.

   _____

Review the Strategies

Choose a strategy
- Write a number sentence.
- Draw a diagram.
- Use logical reasoning.

4. Tammy saw a two-dimensional shape. The shape has 6 sides and 6 angles. Two of the sides are longer than the others. What shape did Tammy see?

_____

5. The sign at the end of David's street is a two-dimensional shape. It has 4 sides and 4 angles. The sides are all the same length. What shape is the sign?

_____

6. Jason was drawing shapes with sidewalk chalk in the park. He drew 3 triangles and then 2 squares. How many angles did he draw?

_____ angles

Chalk is cool!

_____ angles

Name _____

My Homework

Jacob was looking for shapes in the stars. He found one with 4 equal length sides and 4 angles. What shape did Jacob see in the stars?

1 Understand Underline what you know.
Circle what you need to find.

_____

2 Plan How will I solve the problem?

_____

3 Solve Draw a diagram.

Jacob saw a square.

_____

4 Check Is my answer reasonable? Explain.

Underline what you know. Circle what you need to find. Draw a diagram to solve.

Home, sweet home!

1. Maggie drew a house. She drew a square for the bottom. She drew a triangle on top of the square for the roof. Trace the outside of the house. What shape is Maggie's house? Draw it.

_____

2. Billy saw a sign while walking through the park. The sign had no sides and no angles. What shape is the sign?

_____

3. Landon painted a shape that had 4 sides. The shape had 4 angles. What kind of shape did he paint?

_____

Math at Home Describe a shape to your child. Have him or her draw the shape you described and identify the shape.

Name _____

Vocabulary Check

Complete each sentence.

angle	hexagon	pentagon
side	triangle	two-dimensional shape

1. A _____ has 5 sides and 5 angles.

2. A _____ has 3 sides and 3 angles.

3. A _____ has 6 sides and 6 angles.

Concept Check

Circle the shape or shapes that match the name.

4. triangle

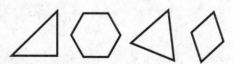

5. pentagon

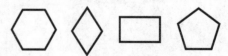

6. hexagon

7. quadrilateral

Write the name of the shape. Circle the shapes that match.

8. _____

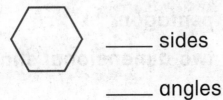

9.  _____

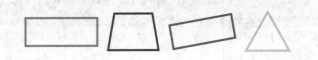

Write how many sides and angles.

10.   _____ sides

                _____ angles

11.   _____ sides

                _____ angles

12.   _____ sides

                _____ angles

13.   _____ sides

                _____ angles

Test Practice

14. Look at the shapes. Mark the shape that does not belong.

    ○   ○   ○   ○

Three-Dimensional Shapes

Lesson 4

ESSENTIAL QUESTION
How do I use shapes and equal parts?

Explore and Explain

Watch ▶   Tools

 Teacher Directions: Look at the picture. Circle the three-dimensional shapes that you see. Identify and describe each shape.

See and Show

A **three-dimensional shape** is a shape with length, width, and height.

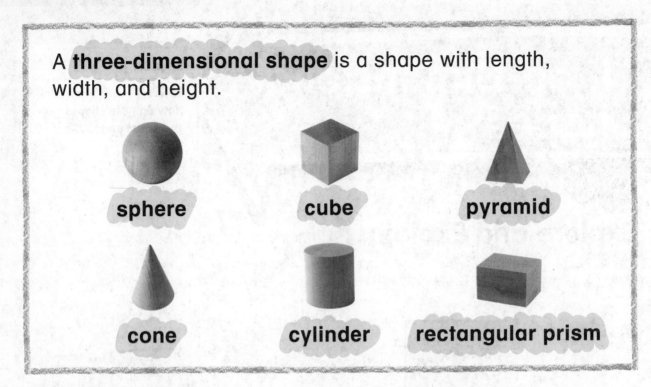

sphere   cube   pyramid

cone   cylinder   rectangular prism

Write the name of the shape. Circle the objects that are the same shape.

1.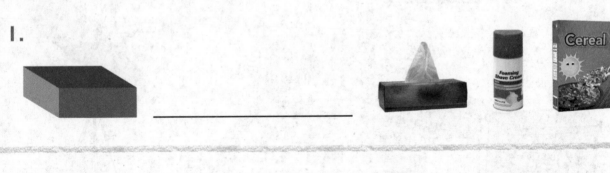

_____

2.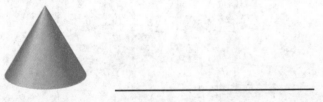

_____

Talk Math Name two objects in your classroom that are the same shape as a rectangular prism.

Name _____

My favorite sphere!

On My Own

Write the name of the shape.
Circle the objects that are the
same shape.

3.   _____

4. 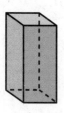 _____

5. 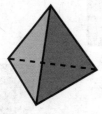 _____

6.   _____

7.   _____

Glue Stick

Problem Solving

8. Jen is wrapping a present that is in a box. The box is square on all sides. What shape is the box Jen is wrapping?

_____

9. I am a three-dimensional shape. I have a circle at the bottom. I have a point at the top. What shape am I?

_____

10. If you stack two cubes together what three-dimensional shape will you make?

_____

Write Math   How can you tell if a shape is three-dimensional?

_____

_____

_____

My Homework

Homework Helper

 eHelp

Need help? connectED.mcgraw-hill.com

A three-dimensional shape has length, width, and height.

sphere   cube   pyramid   cone   cylinder   rectangular prism

Practice

Write the name of the shape. Circle the objects that are the same shape.

1. _____

2.  _____

3. _____

Write the name of each shape. Circle the objects that are the same shape.

4. _____

5. 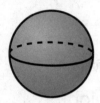 _____

6. I have 6 surfaces. 2 of my surfaces are smaller than the others. I can stand up tall. What shape am I?

_____

Vocabulary Check

Draw lines to match.

7. **cylinder**

8. **rectangular prism**

9. **cube**

10. **cone**

 Math at Home Have your child identify items in your home that match the shapes he or she learned about in this lesson.

Name ....

Faces, Edges, and Vertices

Lesson 5

ESSENTIAL QUESTION
How do I use shapes and equal parts?

Explore and Explain Watch

11
7 •

10
•

4
• 8

3
• 9

6
•

5
• →
1

2
•

We love to climb shapes!

The shape is a _____.

 Teacher Directions: Start at 1. Connect the dots in number order. Write the name of the shape you drew.

See and Show

You can describe three-dimensional shapes by the number of faces, edges, and vertices.

edge → ← face

← vertex

A **face** is a flat surface.

An **edge** is where 2 faces meet.

A **vertex** is where 3 or more faces meet.

Use three-dimensional shapes. Count the faces, edges, and vertices.

Shape	Faces	Edges	Vertices
1. cube	_____	_____	_____
2. rectangular prism	_____	_____	_____
3. pyramid	_____	_____	_____
4. sphere	_____	_____	_____

Talk Math What figure has 6 equal faces?
How do you know?

On My Own

Circle the shapes that match the description.

5. 0 faces, 0 edges, 0 vertices

6. 6 faces, 12 edges, 8 vertices

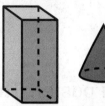

7. 5 faces, 8 edges, 5 vertices

8. 6 equal faces, 12 edges, 8 vertices

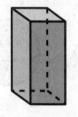

Circle the objects that match the description.

9. 6 faces, 12 edges, 8 vertices

10. 0 faces, 0 edges, 0 vertices

11. 6 faces, 12 edges, 8 vertices

12. 5 faces, 8 edges. 5 vertices

Problem Solving

13. Mindy drew the three shapes below. Circle the shape that has 6 faces and 12 edges.

Take a sphere to the park! We're lots of fun!

14. Ryan has a poster. The poster shows all of the shapes that have less than three faces. Circle the shape that is not on the poster.

HOT Problem Which shape does not belong? Circle it. Explain why it does not belong.

_____

_____

_____

My Homework

Homework Helper **Need help?** connectED.mcgraw-hill.com

Three-dimensional shapes are described
by the number of faces, edges, and vertices.

A face is a flat surface.

An edge is where 2 faces meet.

A vertex is where 3 or more faces meet.

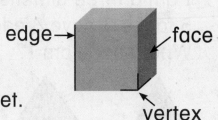

Practice

Circle the shapes or objects that matches the description.

1. 6 faces, 12 edges,
8 vertices

  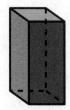

2. 0 faces, 0 edges, 0 vertices

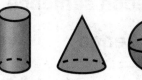

3. 5 faces, 8 edges,
5 vertices

4. 6 faces, 12 edges,
8 vertices

Circle the objects that match the descriptions.

5. 6 faces, 12 edges, 8 vertices

6. 0 faces, 0 edges, 0 vertices

7. I am a three-dimensional shape. I have 5 faces. I have 8 edges and 5 vertices. What shape am I?

 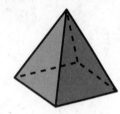

I love parties in the park!

Vocabulary Check

Complete each sentence.

face   **edge**   **vertex**

8. A _____ is a flat surface.

9. A _____ is where 3 or more faces meet.

10. An _____ is where 2 faces meet.

Math at Home Have your child identify real-life objects in your home that have the same shape as one of the shapes learned in this lesson.

770   Chapter 12 • Lesson 5

Name ....

Relate Shapes and Solids

Lesson 6

ESSENTIAL QUESTION
How do I use shapes and equal parts?

Explore and Explain

Yummy cubes!

Teacher Directions: Trace one face of a cube. Identify the shape. Trace the other faces of the cube. Describe the faces of a cube.

See and Show

The faces of three-dimensional
shapes are two-dimensional shapes.

Helpful Hint
A cube has 6 equal faces.
The faces are squares.

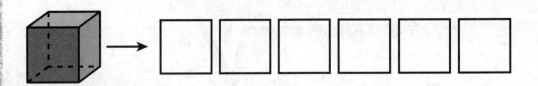

Circle the shapes that make the three-dimensional shape.

1.

2.

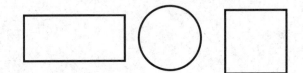

3.

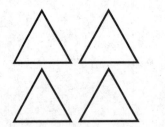

Talk Math   Explain how two-dimensional shapes
and three-dimensional shapes are related.

Name _____

On My Own

Circle the shapes that make the three-dimensional shape.

4.

5.

Circle the shape made by the faces.

6.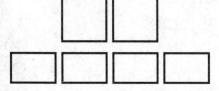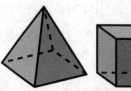

7.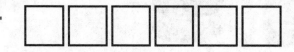

8. Which of these shapes does not have a square as one of its faces?

Problem Solving

9. I have 6 equal faces. I have 8 vertices. What shape am I?

_____

I have only one face!

10. I have no faces and no vertices. What shape am I?

_____

11. Allison wants to trace a circle. Which objects could she use? Circle the objects.

Write Math   Describe the faces that make a pyramid.

_____

_____

_____

Name ....

My Homework

Homework Helper Need help? connectED.mcgraw-hill.com

The faces of three-dimensional shapes
are two-dimensional shapes.

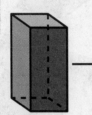   

Helpful Hint
A rectangular prism has
4 rectangles and 2
squares as faces.

Practice

Circle the shapes that make the three-dimensional shape.

1.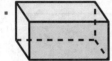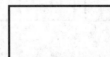

2.

3.

Circle the shape made by the faces.

4.

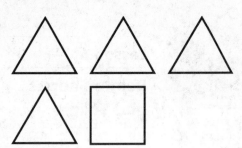 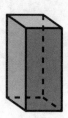

5.

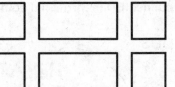 

6. If you put these shapes together, what three-dimensional shape could you make? Write the name of the shape.

Look, we're in a shape!

☐ ☐ ☐ ☐ ☐ ☐ _____

Test Practice

7. Identify the shape that does not belong.

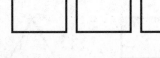

○   ○   ○   ○

Copyright © The McGraw-Hill Companies, Inc.   Ryan McVay/Photodisc/Getty Images

 Math at Home Ask your child to find an object at home that he or she could use to trace a rectangle on a piece of paper. Challenge your child to see if he or she can find something to trace for a circle.

Name ....

Halves, Thirds, and Fourths

Lesson 7

ESSENTIAL QUESTION
How do I use shapes
and equal parts?

Explore and Explain

Teacher Directions: Use square, triangle, and trapezoid pattern blocks to cover
each shape. Trace the blocks to show the shapes you used. Write how
many blocks you used to cover each shape.

See and Show

> You can **partition**, or separate, shapes into equal parts.
>
> Two equal parts or two **halves.**
> Each part is **half of** the whole.
>
> Three equal parts or three **thirds.**
> Each part is a **third of** the whole.
>
> Four equal parts or four **fourths.**
> Each part is a **fourth of** the whole.

Describe the equal parts. Write
two halves, three thirds, **or** *four fourths.*

1.  _____

2. _____

3. 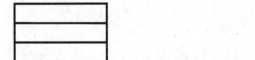 _____

4. _____

Draw lines to partition each shape.

5.

   2 equal parts

6.

   4 equal parts

Talk Math Explain how you can divide a pie so that four people each get an equal part.

Name _____

On My Own

Describe the equal parts. Write
two halves, three thirds, or four fourths.

7. 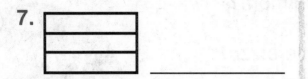 _____

8.  _____

Draw lines to partition each shape.

9.
4 equal parts

10.
2 equal parts

11.
3 equal parts

12.
2 equal parts

**Partition the shape in a different way.
Show the same number of equal parts.**

13.

14.

15.

16.

Problem Solving

What's for lunch?

17. Eva's mom bought a pizza. Eva ate one equal part. Her friend ate one equal part. There was one equal part left for Eva's mom. How much of the pizza was left for Eva's mom?

_____ of the pizza

18. Gracie had a round slice of watermelon. She and her sister shared the slice equally. How much did each girl eat?

_____ of the watermelon slice

19. Sadie is making a picture for her cousin. She folds a piece of paper in half. Then she folds it in half again. She opens the paper. How many equal parts are there?

_____

HOT Problem Show the same number of equal parts in two different ways.

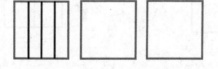

_____

_____

_____

My Homework

Homework Helper **Need help?** connectED.mcgraw-hill.com

You can partition, or separate, shapes into equal parts.

halves

thirds

fourths

Practice

Describe the equal parts. Write *two halves, three thirds, or four fourths.*

1.

_____

2.

_____

3.

_____

4.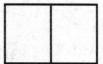

_____

Draw lines to partition each shape.

5.

3 equal parts

6.

2 equal parts

7.

4 equal parts

Sharing halfsies is fun!

8. Nora and Brooke are sharing a sandwich. They each have an equal part. How much of the sandwich does each girl have?

_____ sandwich

Partition the shape in a different way. Show the same number of equal shares.

9.

10.

Vocabulary Check

Vocab abc

Color each shape as described.

11.

one half green

12.

one fourth blue

13.

one third red

 Math at Home Cut your child's food into either halves, thirds, or fourths. Ask him or her to identify how many equal parts you have created.

Name ....

Area

Explore and Explain

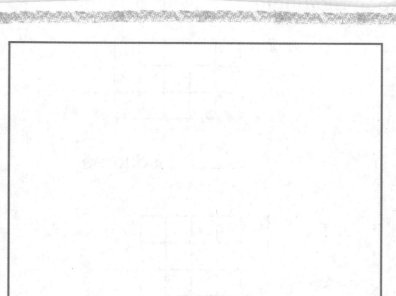

_____

We're for 4-Square!

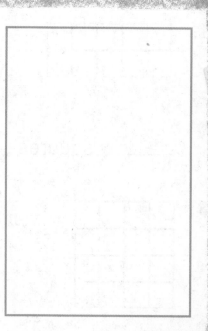

Teacher Directions: Use color tiles to cover each shape. Draw lines to show how the squares fit together to make the shape. Write how many tiles you used.

See and Show

Rectangles can be partitioned into equal-sized squares.
You can count the squares to describe its size.

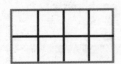 _____8_____ squares

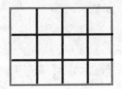 _____12_____ squares

**Count the squares. Write how
many squares make each rectangle.**

1.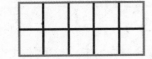

_____ squares

2.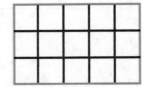

_____ squares

3.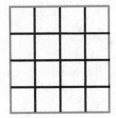

_____ squares

4.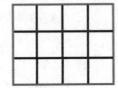

_____ squares

5.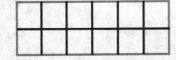

_____ squares

6.

_____ squares

Talk Math Explain how you would partition a
rectangle into 6 equal-sized squares.

Name

_____

On My Own

Count the squares.
Write how many squares make each rectangle.

7.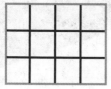

_____ squares

8.

_____ squares

Use color tiles to cover each rectangle.
Write how many tiles were used.

9.

_____ tiles

10.

_____ tiles

Problem Solving

Draw a picture to solve.

11. Addison and Corin each have a group of dominoes. Addison places 4 dominoes in each row. She made 5 rows. Corin places 5 dominoes in each row. She made 3 rows. Who has more dominoes?

_____

12. Larry was placing crackers on a tray. The tray held 4 rows of crackers. There were 16 crackers on the plate in all. How many crackers were in each column?

Yummy!

_____ crackers

Write Math   Explain two ways to partition a rectangle into 4 equal parts.

_____

_____

_____

Name ....

My Homework

Homework Helper Need help? connectED.mcgraw-hill.com

A rectangle can be partitioned into squares to describe its size.

This rectangle is partitioned into 10 squares.

This rectangle is partitioned into 12 squares.

Practice

Count the squares. Write how many squares make each rectangle.

1.

_____ squares

2.

_____ squares

3.

_____ squares

4.

_____ squares

Count the squares. Write how many squares make each rectangle.

Form a straight line and **no** pushing!

5.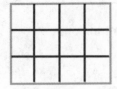

_____ squares

6.

_____ squares

7.

_____ squares

Draw a picture to solve.

8. Jan is cutting a rectangular pan of brownies. She cut them in half and then cut each of those halves in half again. She did the same thing going the other direction. How many brownies does she have?

_____ brownies

Test Practice

9. Choose the rectangle below that is partitioned using the greatest number of squares.

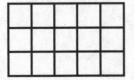   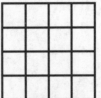 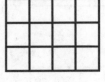

      ○

Math at Home Find an opportunity to have your child help you determine how to cut something you are going to serve such as a casserole, brownies, cake, or rice treats. Together determine how many equal pieces you want and how to cut it.

788   Chapter 12 • Lesson 8

Name ....

My Review

Vocabulary Check

Draw each shape.

1.   hexagon

2.   quadrilateral

Match each word to the correct shape.

3. **cone**

4. **pyramid**

5. **sphere**

6. **cube**

7. **cylinder**

8. **rectangular prism**

Concept Check

Circle the shapes that match the name.

9. rectangle

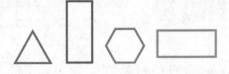

10. pentagon

Circle the shape that matches the description.

11. 3 sides and 3 angles

12. 6 sides and 6 angles

Write the name of the shape. Color the shapes that match.

13.

 _____

Look, it's my twin!

Circle the object that matches the description.

14. 6 faces, 12 edges, 8 vertices

Name _____

Problem Solving

15. Caden's family is having pizza for dinner.
There are 4 people in Caden's family.
Draw lines to show how Caden's
family should cut the pizza so that everyone
gets an equal part.

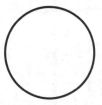

16. Jamie drew a rectangle. He wants to partition
it into equal-sized squares. He starts to partition
the rectangle. Finish partitioning his rectangle.
Write the total number of squares.

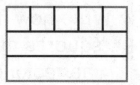   _____ squares

Test Practice

17. Blake has a peanut butter and jelly
sandwich for lunch. He wants to share it
with his 2 friends. Circle the word that tells
how Blake should partition his sandwich.

halves   thirds   fourths   fifths
  ○   ○   ○   ○

Lunchtime!

Reflect

Show the ways you can use shapes and equal parts.

Identify each shape.

 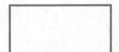

_____   _____

Identify each shape.

 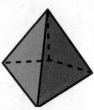

_____   _____

ESSENTIAL QUESTION

How do I use shapes and equal parts?

Partition the shape to match the description.

thirds

Write how many squares make the rectangle.

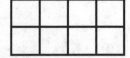

_____ squares

Get the swing of it!

Glossary/Glosario

Vocab
abc

← Go online for the eGlossary.

Aa

English	Spanish/Español
A.M. The hours from midnight until noon.	**a.m.** Las horas que van desde la medianoche hasta el mediodía.
add (addition) Join together sets to find the total or sum. The opposite of *subtract*.	**sumar (adición)** Unir conjuntos para hallar el total o la suma. Lo opuesto de restar.

$$2 + 5 = 7$$

$$2 + 5 = 7$$

addend Any numbers or quantities being added together.

In 2 + 3 = 5, 2 is an addend and 3 is an addend.

$$2 + 3 = 5$$
↑   ↑

sumando Numeros o cantidades que se suman.

En 2 + 3 = 5, 2 es un sumando y 3 es un sumando.

$$2 + 3 = 5$$
↑   ↑

Aa

after Follow in place or time.

5678

6 is just *after* 5

después Que sigue en lugar o en tiempo.

5678

6 está justo *después* del 5

---

analog clock A clock that has an hour hand and a minute hand.

reloj analógico Reloj que tiene una manecilla horaria y un minutero.

---

angle Two sides on a two-dimensional shape meet to form an angle.

ángulo Dos lados de una figura bidimensional se encuentran para formar un ángulo.

---

array Objects displayed in rows and columns.

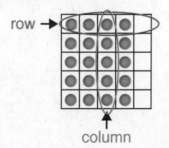

arreglo Objetos organizados en filas y columnas.

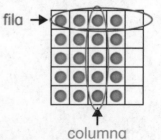

bar graph A graph that uses bars to show data.

gráfica de barras Gráfica que usa barras para ilustrar datos.

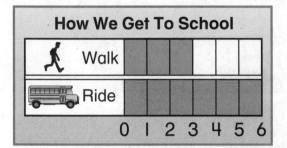

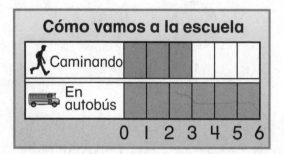

before

5 6 7 8

6 is just *before* 7

antes

5 6 7 8

6 está justo *antes* del 7

between

47 48 49 50

49 is *between* 48 and 50

entre

47 48 49 50

49 está *entre* 48 y 50

cent

I¢   I cent

centavo

I¢   I centavo

Cc

cent sign (¢) The sign used to show cents.

1¢   5¢

signo de centavo (¢) El signo que se usa para mostrar centavos.

1¢   5¢

---

centimeter (cm) A metric unit for measuring length.

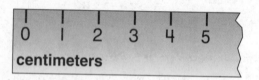

centímetro (cm) Unidad métrica para medir la longitud.

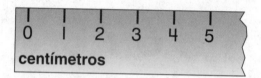

---

circle A closed, round two-dimensional shape.

círculo Bidimensional redonda y cerrada.

---

compare Look at objects, shapes, or numbers and see how they are alike or different.

comparar Observar objetos, formas o números para saber en qué se parecen y en qué se diferencian.

cone A three-dimensional shape that narrows to a point from a circular face.

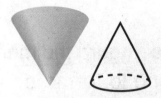

cono Figura tridimensional que se estrecha hasta un punto desde una base circular.

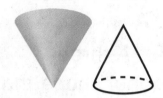

count back On a number line, start at the greater number (5) and count back (3).

$$5 - 3 = 2$$

2 3 4 **5** 6

contar hacia atrás En una recta numérica, comienza en un número mayor (5) y cuenta (3) hacia atrás.

$$5 - 3 = 2$$

2 3 4 **5** 6

count on On a number line, start at the first addend (4) and count on (2).

$$4 + 2 = 6$$

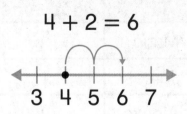

3 4 5 6 7

seguir contando En una recta numérica, comienza en el primer sumando (4) y cuenta (2) hacia delante.

$$4 + 2 = 6$$

3 4 5 6 7

Cc

cube A three-dimensional shape with 6 square faces.

cubo Figura tridimensional con 6 caras cuadradas.

cylinder A three-dimensional shape that is shaped like a can.

cilindro Figura tridimensional que tiena la forma de una lata.

Dd

data Numbers or symbols, sometimes collected from a survey or experiment, that show information. *Data* is plural.

Name	Number of Pets
Mary	3
James	1
Alonzo	4

datos Números o símbolos que se recopilan mediante una encuesta o experimento para mostrar información.

Nombre	Número de mascotas
María	3
James	1
Alonzo	4

day I day = 24 hours
Examples: Sunday,
Monday, Tuesday,
Wednesday, Thursday,
Friday, Saturday

día I día = 24 horas
Ejemplos: domingo, lunes,
martes, miércoles, jueves,
viernes y sábado

difference The answer to a subtraction problem.

$$3 - 1 = 2$$

The difference is 2.

diferencia Resultado de un problema de resta.

$$3 - 1 = 2$$

La diferencia es 2.

digit A symbol used to write numbers. The ten digits are

0, 1, 2, 3, 4, 5, 6, 7, 8, 9.

dígito Símbolo que se utiliza para escribir números. Los diez dígitos son

0, 1, 2, 3, 4, 5, 6, 7, 8, 9.

digital clock A clock that uses only numbers to show time.

reloj digital Reloj que marca la hora solo con números.

Dd

dime dime = 10¢ or 10 cents

head   tail

moneda de 10¢ moneda de diez centavos = 10¢ o 10 centavos

cara   cruz

dollar one dollar =100¢ or 100 cents. It can also be written as $1.00 or $1.

front

back

dólar un dólar = 100¢ o 100 centavos. También se puede escribir $1.00.

frente

revés

dollar sign ($) The sign used to show dollars.

one dollar = $1 or $1.00

signo de dólar ($) Símbolo que se usa para mostrar dólares.

un dólar = $1 o $1.00

doubles (and near doubles) Two addends that are the same number.

$6 + 6 = 12$ ← doubles

$6 + 7 = 13$ ← near doubles

dobles (y casi dobles) Dos sumandos que son el mismo número.

$6 + 6 = 12$ ← dobles

$6 + 7 = 13$ ← casi dobles

Ee

edge The line segment where two *faces* of a three-dimensional shape meet.

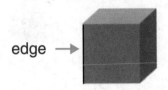

arista Segmento de recta donde se encuentran dos caras de una figura tridimensional.

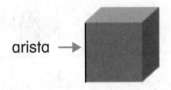

equal groups Each group has the same number of objects.

There are four equal groups of counters.

grupos iguales Cada grupo tiene el mismo número de objetos.

Hay cuatro grupos iguales de fichas.

equal parts Each part is the same size.

This sandwich is cut into 2 equal parts.

partes iguales Cada parte es del mismo tamaño.

Este sándwich está cortado en 2 partes iguales.

Ee

equal to =

6 = 6

6 is *equal to* or the same as 6.

igual a =

6 = 6

6 es *igual* o lo mismo que 6.

estimate Find a number close to an exact amount.

47 + 22 rounds to 50 + 20.
The estimate is 70.

estimar Hallar un número cercano a la cantidad exacta.

47 + 22 se redondea a 50 + 20.
La estimación es 70.

even number Numbers that end with 0, 2, 4, 6, 8.

número par Los números que terminan en 0, 2, 4, 6, 8.

expanded form
The representation of a number as a sum that shows the value of each digit. Sometimes called *expanded notation*.

536 is written as 500 + 30 + 6.

forma desarrollada
La representación de un número como suma que muestra el valor de cada dígito. También se llama *notación desarrollada*.

536 se escribe como 500 + 30 + 6.

face The flat part of a three-dimensional shape.

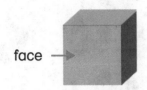

A square is a face of a cube.

cara La parte plana de una figura tridimensional.

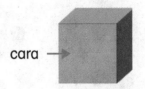

El cuadrado es la cara de un cubo.

fact family Addition and subtraction sentences that use the same numbers.

$$6 + 7 = 13 \qquad 13 - 7 = 6$$
$$7 + 6 = 13 \qquad 13 - 6 = 7$$

familia de operaciones Enunciados de suma y resta los cuales tienen los mismos números.

$$6 + 7 = 13 \qquad 13 - 7 = 6$$
$$7 + 6 = 13 \qquad 13 - 6 = 7$$

foot (ft) A customary unit for measuring length. Plural is feet.

1 foot = 12 inches

pie Unidad usual para medir longitud.

1 pie = 12 pulgadas

fourths Four equal parts of a whole. Each part is a fourth, or a quarter of the whole.

cuartos Cuatro partes iguales de un todo. Cada parte es un cuarto, o la cuarta parte del todo.

greater than >

7 > 2

7 is greater than 2.

mayor que >

7 > 2

7 es mayor que 2.

group A set of objects.

I group of 4

grupo Conjunto o grupo de objetos.

I grupo de 4

half hour (or half past)
A unit to measure time.
Sometimes called *half past*
or *half past the hour*.

a half hour = 30 minutes

media hora (o y media)
Unidad para medir tiempo.
A veces se dice *hora y
media.*

media hora = 30 minutos

halves Two equal parts of
a whole. Each part is a half
of the whole.

mitades Dos partes iguales
de un todo. Cada parte es
la mitad de un todo.

hexagon A 2-dimensional
shape that has six sides.

hexágono Figura
bidimensional que tiene seis
lados.

hour A unit to measure time.

I hour = 60 minutes

hora Unidad para medir tiempo.

I hora = 60 minutos

hour hand The hand on a clock that tells the hour. It is the shorter hand.

hour hand

manecilla horaria Manecilla del reloj que indica la hora. Es la más corta.

manecilla horaria

hundreds The numbers in the range of 100–999. It is the place value of a number.

365

3 is in the hundreds place.
6 is in the tens place.
5 is in the ones place.

centenas Los números en el rango de 100 a 999. Es el valor posicional de un número.

365

3 está en el lugar de las centenas.
6 está en el lugar de las decenas.
5 está en el lugar de las unidades.

inch (in) A customary unit for measuring length. The plural is *inches*.

12 inches = 1 foot

pulgada (pulg) Unidad usual para medir longitud.

12 pulgadas = 1 pie

inverse Operations that are opposite of each other.

Addition and subtraction are inverse or opposite operations.

operaciones inversas Operaciones que se oponen una a otra.

La suma y la resta son operaciones inversas u opuestas.

Kk

key Tells what (or how many) each symbol stands for.

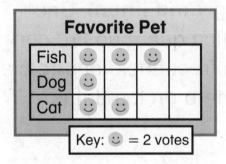

clave Indica qué o cuánto representa cada símbolo.

Animal doméstico favorito			
Pez	☺	☺	☺
Perro	☺		
Gato	☺	☺	

Clave: ☺ = 2 votos

length How long or how far away something is.

length

longitud El largo de algo o lo lejos que está.

longitud

less than <

4 < 7

4 is less than 7.

menor que <

4 < 7

4 es menor que 7.

line plot A graph that shows how often a certain number occurs in data.

diagrama lineal Una gráfica que muestra con qué frecuencia ocurre cierto número en los datos.

measure To find the length, height, weight, capacity, or temperature using standard or nonstandard units.

medir Hallar la longitud, altura, peso, capacidad o temperatura usando unidades estándares o no estándares.

meter (m) A metric unit for measuring length.

I meter = 100 centimeters

metro (m) Unidad métrica para medir longitud.

I metro = 100 centímetros

minute (min) A unit to measure time. Each tick mark is one minute.

I minute = 60 seconds

minuto (min) Unidad para medir tiempo. Cada marca es un minuto.

I minuto = 60 segundos

minute hand The longer hand on a clock that tells the minutes.

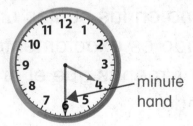

minutero La manecilla más larga del reloj. Indica los minutos.

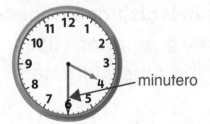

Mm

missing addend The missing number in a number sentence that makes the number sentence true.

$$9 + \boxed{} = 16$$

The missing addend is 7.

sumando desconocido El número desconocido en un enunciado numérico que hace que este sea verdadero.

$$9 + \boxed{} = 16$$

El sumando desconocido es 7.

month A unit of time.
12 months = 1 year

April						
Sunday	Monday	Tuesday	Wednesday	Thursday	Friday	Saturday
		1	2	3	4	5
6	7	8	9	10	11	12
13	14	15	16	17	18	19
20	21	22	23	24	25	26
27	28	29	30			

This is the month of April.

mes Unidad de tiempo.
12 mesas = 1 año

Abril						
domingo	lunes	martes	miércoles	jueves	viernes	sábado
		1	2	3	4	5
6	7	8	9	10	11	12
13	14	15	16	17	18	19
20	21	22	23	24	25	26
27	28	29	30			

Este es el mes de abril.

Nn

near doubles Addition facts in which one addend is exactly 1 more or 1 less than the other addend.

casi dobles Operaciones de suma en las cuales un sumando es exactamente 1 más o 1 menos que el otro sumando.

nickel nickel = 5¢ or 5 cents

head   tail

moneda de 5¢ moneda de cinco centavos = 5¢ o 5 centavos

cara   cruz

number line A line with number labels.

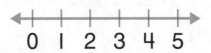

recta numérica Recta con marcas de números.

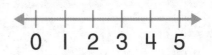

o'clock At the beginning of the hour.

It is 7 o'clock.

en punto El momento en que comienza cada hora.

Son las 7 en punto.

odd number Numbers that end with 1, 3, 5, 7, 9.

números impares Los números que terminan en 1, 3, 5, 7, 9.

Oo

ones The numbers in the range of 0-9. A place value of a number.

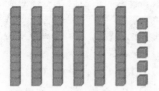

65

5 is in the ones place.

unidades Los números en el rango de 0 a 9. Valor posicional de un número.

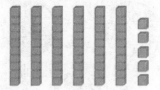

65

El 5 está en el lugar de las unidades.

order

1, 3, 6, 8, 10

These numbers are in order from least to greatest.

orden

1, 3, 6, 8, 10

Estos números están en orden de menor a mayor.

P.M. The hours from noon until midnight.

p.m. Las horas que van desde el mediodía hasta la medianoche.

parallelogram A two-dimensional shape that has four sides. Each pair of opposite sides is equal and parallel.

paralelogramo Figura bidimensional que tiene cuatro lados. Cada par de lados opuestos son iguales y paralelos.

partition To divide or "break up."

separar Dividir o desunir.

pattern An order that a set of objects or numbers follows over and over.

pattern unit

patrón Orden que sigue continuamente un conjunto de objetos o números insertar punto.

unidad de patrón

penny penny = 1¢ or 1 cent

head   tail

moneda de 1¢ moneda de un centavo = 1¢ o 1 centavo

cara   escudo

pentagon A polygon with five sides.

pentágono Polígono de cinco lados.

picture graph A graph that has different pictures to show information collected.

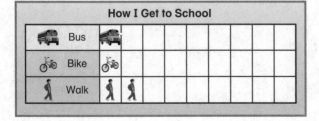

gráfica con imágenes Gráfica que tiene diferentes imágenes para ilustrar la información recopilada.

place value The value given to a *digit* by its place in a number.

1,365

1 is in the thousands place.
3 is in the hundreds place.
6 is in the tens place.
5 is in the ones place.

valor posicional El valor dado a un *dígito* según su posición en un número.

1,365

1 está en el lugar de los millares.
3 está en el lugar de las centenas.
6 está en el lugar de las decenas.
5 está en el lugar de las unidades.

pyramid A three-dimensional shape with a polygon as a base and other faces that are triangles.

pirámide Figura tridimensional con un polígono como base y otras caras que son triángulos.

Qq

quadrilateral A shape that has 4 sides and 4 angles.

cuadrilátero Figura con 4 lados y 4 ángulos.

quarter quarter = 25¢ or 25 cents

head   tail

moneda de 25¢ moneda de 25 centavos = 25¢ o 25 centavos

cara   cruz

quarter hour A quarter hour is 15 minutes. Sometimes called *quarter past* or *quarter til*.

cuarto de hora Un cuarto de hora es 15 minutos. A veces se dice hora y cuarto.

rectangle A plane shape with four sides and four corners.

rectángulo Figura plana con cuatro lados y cuatro esquinas.

rectangular prism
A three-dimensional shape
with 6 faces that are
rectangles.

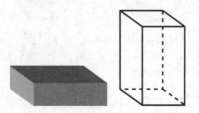

prisma rectangular Figura
tridimensional con 6 caras
que son rectángulos.

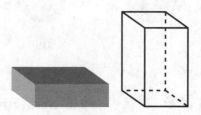

regroup Take apart a
number to write it in a new
way.

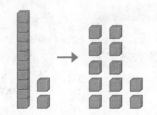

1 ten + 2 ones becomes
12 ones.

reagrupar Separar un
número para escribirlo de
una nueva forma.

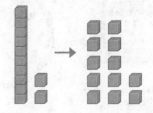

1 decena + 2 unidades se
convierten en 12 unidades.

related fact(s) Basic facts
using the same numbers.
Sometimes called a *fact
family*.

$$4 + 1 = 5 \quad 5 - 4 = 1$$
$$1 + 4 = 5 \quad 5 - 1 = 4$$

operaciones relacionadas
Operaciones básicas en las
que se usan los mismos
números. También se
llaman *familias de
operaciones*.

$$4 + 1 = 5 \quad 5 - 4 = 1$$
$$1 + 4 = 5 \quad 5 - 1 = 4$$

repeated addition To use the same addend over and over.

suma repetida Usar el mismo sumando una y otra vez.

rhombus A shape with 4 sides of the same length.

rombo Paralelogramo con cuatro lados de la misma longitud.

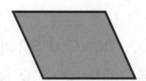

round Change the *value* of a number to one that is easier to work with.

redondear Cambiar el *valor* de un número a uno con el que es más fácil trabajar.

24 rounded to the nearest ten is 20.

24 redondeado a la decena más cercana es 20.

side One of the line segments that make up a shape.

lado Uno de los segmentos de recta que componen una figura.

A pentagon has five sides.

El pentágono tiene cinco lados.

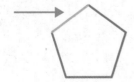

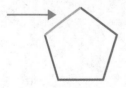

Ss

---

skip count Count objects in equal groups of two or more.

2, 4, 6, 8, 10

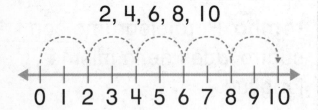

contar salteado Contar objetos en grupos iguales de dos o más.

2, 4, 6, 8, 10

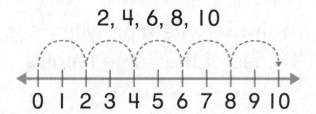

---

slide To move a shape in any direction to another place.

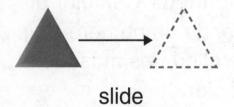

slide

deslizar Traslador una figura a una nueva posición.

deslizar

---

sphere A three-dimensional shape that has the shape of a round ball.

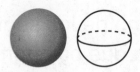

esfera Figura tridimensional que tiene la forma de una pelota redonda.

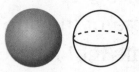

---

square A two-dimensional shape that has four equal sides. Also a rectangle.

cuadrado Figura bidimensional que tiene cuatro lados iguales. También es un rectángulo.

subtract (subtraction) Take away, take apart, separate, or find the difference between two sets. The opposite of *add*.

$$5 - 5 = 0$$

restar (sustracción) Eliminar, quitar, separar o hallar la diferencia entre dos conjuntos. Lo opuesto de *sumar*.

$$5 - 5 = 0$$

sum The answer to an addition problem.

$$2 + 4 = 6$$

suma Resultado de la operación de sumar.

$$2 + 4 = 6$$

survey Collect data by asking people the same question.

Favorite Animal	
Dog	$\bcancel{||||}\,|$
Cat	$\bcancel{||||}$

This survey shows favorite animals.

encuesta Recopilar datos al hacer las mismas preguntas a un grupo de personas.

Animal favorito	
Perro	$\bcancel{||||}\,|$
Gato	$\bcancel{||||}$

Esta encuesta muestra los animales favoritos.

symbol A letter or figure that stands for something.

This symbol means to add.

símbolo Letra o figura que representa algo.

Este símbolo significa sumar.

Tt

tally marks A mark used to record data collected in a survey.

$\bcancel{||||}\,||$

tally marks

marca de conteo Marca que se utiliza para registrar los datos recopilados en una encuesta.

$\bcancel{||||}\,||$

marcas de conteo

tens The numbers in the range of 10–99. It is the place value of a number.

65

6 is in the tens place.
5 is in the ones place.

decenas Los números en el rango de 10 a 99. Es el valor posicional de un número.

65

6 está en el lugar de las decenas.
5 está en el lugar de las unidades.

thirds Three equal parts.

tercios Tres partes iguales.

thousand(s) The numbers in the range of 1,000–9,999. It is the place value of a number.

1,365

1 is in the thousands place.
3 is in the hundreds place.
6 is in the tens place.
5 is in the ones place.

millar(es) Los números en el rango de 1,000 a 9,999. Es el valor posicional de un número.

1,365

1 está en el lugar de los millares.
3 está en el lugar de las centenas.
6 está en el lugar de las decenas.
5 está en el lugar de las unidades.

three-dimensional shape A shape that has length, width, and height.

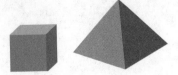

figura tridimensional Que tiene tres dimensiones: largo, ancho y alto.

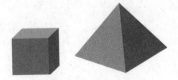

trapezoid A two-dimensional shape with four sides and only two opposite sides that are parallel.

trapecio Figura bidimensional de cuatro lados con solo dos lados opuestos que son paralelos.

triangle A three-dimensional shape with three sides and three angles.

triángulo Figura tridimensional con tres lados y tres ángulos.

two-dimensional shape The outline of a shape - such as a triangle, square, or rectangle - that has only *length* and *width*.

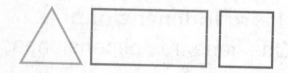

figura bidimensional Contorno de una figura, como un triángulo, un cuadrado o un rectángulo, que solo tiene *largo* y *ancho*.

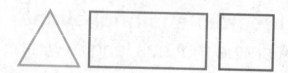

vertex

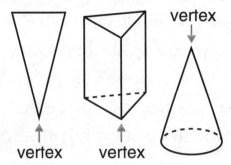

vertex

vertex   vertex

vértice

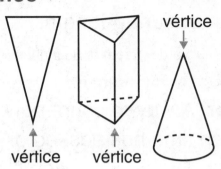

vértice

vértice   vértice

Ww

week A part of a calendar.
I week = 7 days

semana Parte de un calendario
una semana = 7 días

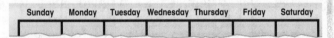

Sunday	Monday	Tuesday	Wednesday	Thursday	Friday	Saturday

domingo	lunes	martes	miércoles	jueves	viernes	sábado

whole The entire amount or object.

el todo La cantidad total o el objeto completo.

yard (yd) A customary unit for measuring length.

1 yard = 3 feet or 36 inches

yarda Unidad usual para medir la longitud.

1 yarda = 3 pies o 36 pulgadas

year A way to count how much time has passed or will pass. 1 year = 12 months

año Un período insertar punto, un año = 12 meses

January	February	March
enero	febrero	marzo

April	May	June
abril	mayo	junio

July	August	September
julio	agosto	septiembre

October	November	December
octubre	noviembre	diciembre

Name

Work Mat 3: Number Lines

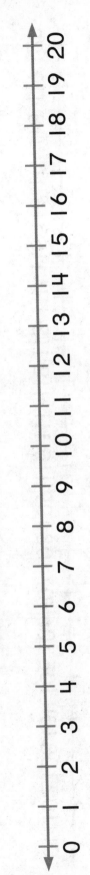

0 1 2 3 4 5 6 7 8 9 10 11 12 13 14 15 16 17 18 19 20

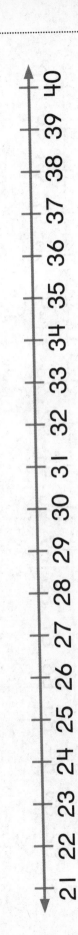

21 22 23 24 25 26 27 28 29 30 31 32 33 34 35 36 37 38 39 40

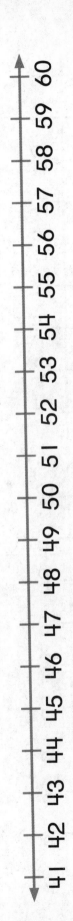

41 42 43 44 45 46 47 48 49 50 51 52 53 54 55 56 57 58 59 60

Work Mat 3: Number Lines   WM1

Work Mat 4: Number Lines

61 62 63 64 65 66 67 68 69 70 71 72 73 74 75 76 77 78 79 80

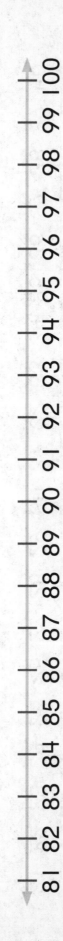

81 82 83 84 85 86 87 88 89 90 91 92 93 94 95 96 97 98 99 100

101 102 103 104 105 106 107 108 109 110 111 112 113 114 115 116 117 118 119 120